Contents

earthworks
plus
John Widdowson

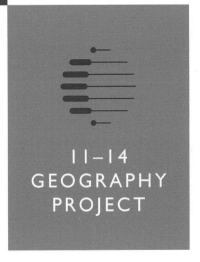

11–14
GEOGRAPHY
PROJECT

JOHN MURRAY

Author's acknowledgements

The author would like to thank the following people who have contributed to *Earthworks Plus*: Catherine Hurst; Dale Banham for his advice on implementing the literacy strategy; Alan Widdowson for research in China; and all the people who agreed to be photographed and interviewed and who appear in the book. Thanks also to Linda Hack of St Mildred's Primary School, Stoke-on-Trent for her contribution to the Antarctica material.

The units in the Earthworks books are made up of the following elements:

GROUNDWORK – an introduction to the unit, based on your everyday experience of geography.

FRAMEWORK – covers all the key geographical ideas that you need to understand in the unit.

BUILDING BLOCKS – two or three geographical investigations, based on the *Framework* ideas, using real places.

DIGGING DEEPER – an in-depth look at a topical issue, to take your geography that little bit further.

  These symbols show the best opportunities for developing your ICT, literacy and numeracy skills alongside your geography work.

First published 2002
by John Murray (Publishers) Ltd
50 Albemarle Street
London W1S 4BD

Layouts by Amanda Easter
Artwork by Oxford Designers & Illustrators Ltd
Cover design by John Townson/Creation
Typeset in Sabon by Wearset Ltd, Boldon, Tyne and Wear
Colour separations by Colourscript, Mildenhall, Suffolk
Printed and bound by G. Canale, Torino, Italy

A catalogue entry for this book is available from the British Library.
ISBN 0 7195 7567 2
Teacher's Resource Book 0 7195 7568 0

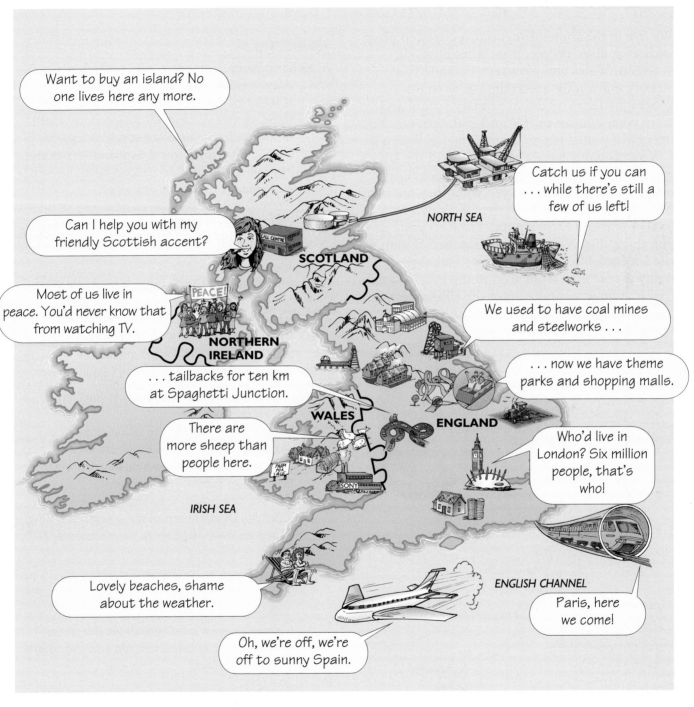

Want to buy an island? No one lives here any more.

Catch us if you can . . . while there's still a few of us left!

NORTH SEA

Can I help you with my friendly Scottish accent?

SCOTLAND

Most of us live in peace. You'd never know that from watching TV.

PEACE!

NORTHERN IRELAND

We used to have coal mines and steelworks . . .

. . . now we have theme parks and shopping malls.

. . . tailbacks for ten km at Spaghetti Junction.

WALES

ENGLAND

There are more sheep than people here.

FARM FOR SALE

SONY

Who'd live in London? Six million people, that's who!

IRISH SEA

Lovely beaches, shame about the weather.

ENGLISH CHANNEL

Paris, here we come!

Oh, we're off, we're off to sunny Spain.

- Which part of the United Kingdom do you live in?
- What do you think other parts of the UK are like?
- Do you think that the comments on this map are fair, or not?
- What would you like to add to the map? (Or what would you take away?)

GROUNDWORK

1.1 Send us a postcard!

We live in the United Kingdom, a crowded place with 59 million people. It includes England, Scotland, Wales and Northern Ireland. You already know a lot about the UK from the geography that you did in primary school and from what you see on TV and in the papers. You may have visited places around the UK yourself.

You probably have a **mental map** of the UK. A mental map is a kind of picture in a person's mind of where places are and what they are like. Map A is a mental map of the UK drawn by a twelve-year-old French pupil. How accurate do you think it is? What would your mental map of the UK look like? In *Earthworks Plus* there are many case studies from around the UK. You may find that your mental map of the UK has changed by the end of the book.

Activities

1 a) Draw your own mental map of the UK. Don't use any other maps to help you. (Your teacher may get you to do this without the book, so that you are not tempted to cheat!) Draw the coastline as accurately as you can. Add borders for England, Scotland, Wales and Northern Ireland. Add features such as rivers, mountains, cities and motorways.

b) Compare your map with an atlas map of the UK. How accurate is your map? How would you change it if you were to draw it again?

2 Look at map A.

a) If you did not already know, how could you tell that it was drawn by a French pupil? Give three different pieces of evidence.

b) What else does this map tell you about the pupil? Do you think he has visited the UK? Give more evidence to support your answer.

3 [L] Work with a partner. Look at postcards B–G on page 3.

a) Suggest which part of the UK each postcard was sent from. You could use the map on page 1 to help you. Give evidence from each postcard to support your ideas.

b) Choose one postcard to analyse more carefully. Discuss each of the questions below with a partner. Then write your answers down.

- What does it tell you about the place?
- What can you guess about the place (including its location)?
- What doesn't it tell you about the place?
- What other questions would you like to ask about the place?

Your teacher may give you a sheet that will give more information about each of the places shown in the postcards.

c) Write a message to a friend to go on the back of the postcard. Imagine that you are on holiday at the place in the postcard.

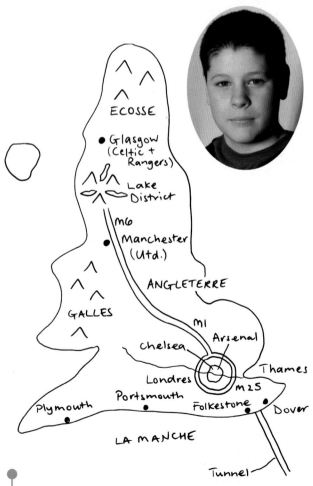

A | A mental map of the UK drawn by a French pupil

B

E

C

F

D

G

1.2 Changing borders

A country is an area of land surrounded by a **border** or a boundary. Borders are often marked by natural features such as rivers or mountain ranges. You can find features like these on a **physical map**. Sometimes the border between two countries is simply a line drawn on a map. In real life on the ground this may be invisible. You can find border lines on a **political map** (see map A).

Borders can lead to problems! They are often the cause of conflict between countries, and even wars. As a result, borders change. Over the past 2,000 years the borders of the countries in the British Isles have changed many times as you can see in source B.

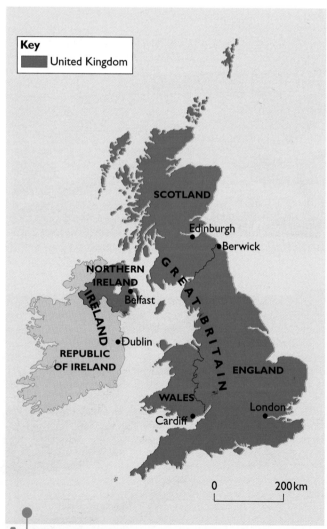

A The British Isles has two main islands – Great Britain and Ireland. The United Kingdom is coloured in red.

AD200

AD900

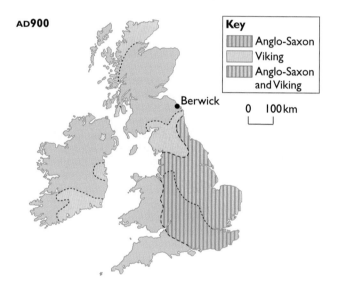

AD1300

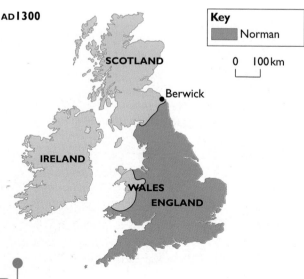

B Changing borders in the British Isles over the past 2,000 years.

	Population (millions)	Area (km²)
England	49.7	130,422
Scotland	5.1	78,133
Wales	2.9	20,779
Northern Ireland	1.7	13,576
Republic of Ireland	3.6	70,273

D Population and area of the countries in the British Isles

C Hadrian's Wall was built nearly 2,000 years ago by the Roman Emperor, Hadrian, to keep the Scots out of England. It marked the northern border of the Roman Empire.

Activities

1 Look at map A. It shows the British Isles, the United Kingdom and Great Britain.
   a) Make a large copy of the table below.
   b) Under the correct headings in the table, list each of the countries: England, Scotland, Wales, Northern Ireland and Republic of Ireland. Some will appear two or three times.

British Isles	United Kingdom	Great Britain

2 Look at photo C.
   a) Using evidence from the photo, suggest why the Romans built the wall here.
   b) Locate Hadrian's Wall on the first map in source B. How has the border between England and Scotland changed since then?
   c) Historically, the town of Berwick has been in both England and Scotland. Explain how this is possible. Which country is it in today?
   d) Locate your own area on the maps in source B. Which country is it in? Has it ever been part of another country? Which groups of people have ruled that area?

3 Look at table D.
   Use the data in the table to compare the five countries in the British Isles. Write two paragraphs to compare:
   a) population
   b) area.
   The first paragraph could start like this:

   The country in the British Isles with the largest population is _____, with _____ million people. The next largest population . . .

4 You are going to reorganise the data in table D to compare the British Isles, the United Kingdom and Great Britain.
   a) Make a large copy of the table below.
   b) Look back to the table that you drew for activity 1. Group the countries in the same way and add the figures for population and area from table D. Write the totals in the table you have drawn.

	Population (millions)	Area (km²)
British Isles		
United Kingdom		
Great Britain		

FRAMEWORK

1.3 On the political map

Political maps show artificial areas that have been created by people. An atlas map showing countries and cities is an example of a political map. Each country has its own **government** located in the **capital city**. The United Kingdom's government is in London at the Houses of Parliament (see photo A).

Political maps don't always show countries. Map B is a political map that shows **counties** in the UK. Large urban areas, including all the major cities, are known as **metropolitan counties**. Because they have such large populations, cities are divided into separate **boroughs**. Greater London alone has 33 boroughs. Map C shows how the West Midlands Metropolitan County is divided. Each county, or borough, is under the control of its own council, or **local government**.

A The Houses of Parliament in London

MC = metropolitan county

B Counties and metropolitan counties in the UK. Smaller unlabelled areas are the unitary authorities (areas with their own local government).

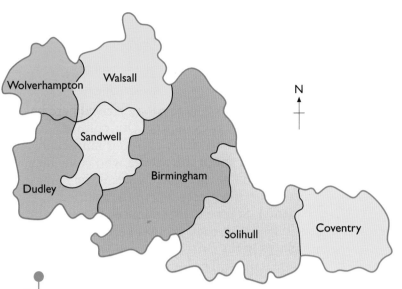

C West Midlands Metropolitan County is divided into seven authorities.

D Birmingham town hall, where the city council meets

Activities

1 Look at the list of UK place names in the box below.

England	Belfast	Aberystwyth	Lancashire
Highland	Taunton	Ceredigion	Shrewsbury
Scotland	Preston	Kent	Inverness
Maidstone	Devon	Cambridge	Merseyside
Birmingham	Wales	Antrim	Somerset
Londonderry	Liverpool	Shropshire	
Northern Ireland	Cambridgeshire		
Coventry	Exeter	West Midlands	

a) With the help of map B and your atlas, classify the list into three groups: cities, counties and countries.

b) Sort them into sets of three. Each set should start with a city, then the county that it is in and, finally, the country they are both in. For example, Exeter – Devon – England. You can use the same county or country more than once.

2 Look at the list of jobs in the box below. For each one, suggest whether they would best be done by national or local government, or both together. In each case, give a reason for your choice.

| street cleaning   transport planning |
| running a health service   providing a water supply |
| nuclear power generation   immigration control |
| creating new jobs   making new laws |
| running a library service |

People vote in elections for the type of government they want. In the UK we elect a new national government for up to five years. The country is divided into 659 **constituencies**. People in each constituency elect an **MP** (Member of Parliament) to represent them. A similar system works for local government too. People living in each district elect local councillors to represent them on the council.

Governments decide how the country should be run and how the taxes that we pay should be spent. Responsibilities are divided between national government and local government. Some jobs – for example, organising the country's defence – are best done nationally, while others – for example, running schools – are best done locally.

Homework

3 a) Which borough/city/county do you live in? Find out the name of the constituency you live in too.

b) Find out the names of the politicians who represent your area in national and local government. Which political parties do they represent?

ICT You could do your research on the internet using the website: www.explore.parliament.uk.

1.4 We are all British

Throughout history, wave after wave of people have come to the British Isles. Some came to conquer, others just settled here quietly. Map A shows the main groups of people that arrived before 1066. How many of these groups have you studied in history?

If each of us could trace our family back far enough we would find that our ancestors all **migrated** here. The next time that you hear someone complain that the country is being swamped with **immigrants**, or that there are too many **refugees**, you could remind them of this! Alternatively, you could read them poem B.

Each group of people that has arrived here has left their mark on the country. Our culture, language, towns and cities, even our landscape, all give us clues about the people who came here. One of the main geographical clues is the names people gave to the places where they settled (**settlements**). Source C lists some of the place names linked with each group of settlers. How many places in the British Isles can you think of with these names? Try looking in an atlas to see how many more you can find.

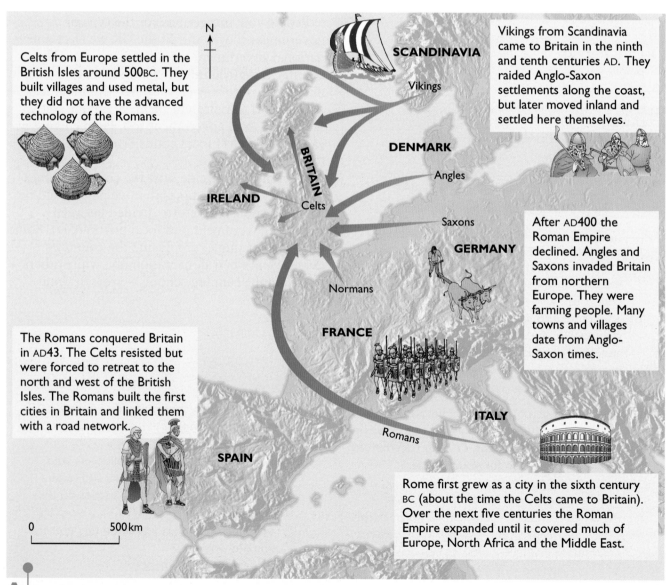

Celts from Europe settled in the British Isles around 500BC. They built villages and used metal, but they did not have the advanced technology of the Romans.

Vikings from Scandinavia came to Britain in the ninth and tenth centuries AD. They raided Anglo-Saxon settlements along the coast, but later moved inland and settled here themselves.

After AD400 the Roman Empire declined. Angles and Saxons invaded Britain from northern Europe. They were farming people. Many towns and villages date from Anglo-Saxon times.

The Romans conquered Britain in AD43. The Celts resisted but were forced to retreat to the north and west of the British Isles. The Romans built the first cities in Britain and linked them with a road network.

Rome first grew as a city in the sixth century BC (about the time the Celts came to Britain). Over the next five centuries the Roman Empire expanded until it covered much of Europe, North Africa and the Middle East.

0   500 km

A People who came to the British Isles before 1066

The British

Serves 60 million

Take some Picts, Celts and Silures
And let them settle,
Then overrun them with Roman conquerors.

Remove the Romans after approximately 400 years
Add lots of Norman French to some
Angles, Saxons, Jutes and Vikings, then stir vigorously.

Mix some hot Chileans, cool Jamaicans, Dominicans,
Trinidadians and Bajans with some Ethiopians,
Chinese, Vietnamese and Sudanese.

Then take a blend of Somalians, Sri Lankans, Nigerians
And Pakistanis,
Combine with some Guyanese
And turn up the heat.

Sprinkle some fresh Indians, Malaysians, Bosnians,
Iraqis and Bangladeshis together with some
Afghans, Spanish, Turkish, Kurdish, Japanese
And Palestinians
Then add to the melting pot.

Leave the ingredients to simmer.

As they mix and blend allow
  their languages to
flourish
Binding them together with
  English.

Allow time to be cool.

**Add some unity,
  understanding and respect for
  the future
Serve with justice
And enjoy.**

*Note: All the ingredients are equally important. Treating one
ingredient better than another will leave a bitter, unpleasant taste.*

*Warning: An unequal spread of justice will damage the people
and cause pain.*

Give justice and equality to all.

B | The British – a poem by Benjamin Zephaniah (see photo)

Celtic names

beginning with:	mean:
Aber-	river mouth
Pen-	hill or headland
Pol-	stream

Roman names

ending with:	mean:
-chester	castle
-caster	castle
-port	harbour

Anglo-Saxon names

ending with:	mean:
-ford	river crossing
-ton	settlement
-ing	land belonging to a person

Viking names

ending with:	mean:
-by	settlement
-thorpe	farm
-ey	island

C | Place names in Britain and their meanings

Activities

I You are going to make a table about early settlement in the British Isles.
      Work in a group of four. Each person could research one of the
groups of people that came to the British Isles.
   a) Draw a table like the one below to fill a page in your workbook.

Group of people	Where did they come from?	When did they come?	What are their settlements called?	Where did they settle?

   b) Study map A. List the groups of people in the first column of your
        table in the order in which they came. Complete the next two
        columns using information on the map.
   c) Look at box C. Use an atlas map of the UK to find five place
        names given by each group. List them in the fourth column.
   d) Where in the UK are the settlements located? Describe their
        distribution in the fifth column.

2   Read poem B.

   a) Is the poet in favour of immigration? Which lines indicate this?
   b) The term 'melting pot' is sometimes used to describe our
        country. What do you think it means?
   c) Why do you think the **bold** lines of the poem are so important?

BUILDING BLOCKS

In this Building Block you will read the story of one refugee family and consider some myths and facts about immigration into the UK.

1.5

The UK welcomes refugees – OK?

Zhao (pronounced *jow*) is in his first year at Astley High School in Manchester. He has spent all his school life in this country and remembers very little about China. That isn't surprising since he was only six when he left China to start a new life here. But the UK government wants to send Zhao and the rest of his family back to China. His school has got involved in a campaign to let Zhao and his brother and sister stay in the UK.

The problems really began for Zhao's family in 1989. That was the year that Zhao was born. It was also the year when students in China organised massive protests about the lack of freedom and **democracy** in China. (Democracy is the system of government that we have in the UK that gives everyone the right to vote.) In June that year thousands of angry demonstrators gathered in Tiananmen Square in the centre of the capital, Beijing. The Chinese government has a reputation for the harsh treatment of its critics, and the world watched and waited to see how it would respond. They did not have to wait long. A few days later army tanks rolled into the square and turned their guns on the students. No one knows the precise figure, but hundreds were killed. Around China thousands more students were put in jail or went into hiding from the government.

At the time of the Tiananmen Square massacre Zhao's father, Lian Hu Su, was a chef in Beijing. Like many Chinese people, he was sympathetic to the demands of the students. He sheltered students who had escaped the massacre and, in doing so, put his own life and freedom in danger. Fearing that the authorities would eventually catch up with him he fled China and eventually ended up in the UK as a refugee (see box B on page 11). For many years this country has had a reputation for its tolerance of refugees so he guessed that this would be a safe place to live. He made plans for the rest of his family to join him. Zhao arrived in the UK in 1995 with his mother, Xiao Fang, and sister and brother, Miao and Jing.

They settled in Manchester where Zhao started to attend Lyndhurst Primary School. His brother and sister went to the neighbouring secondary school, Astley High.

Like many immigrants they were unable to speak English when they arrived. But Zhao learned quickly and by the time that he left primary school he was near the top of his class in all his subjects. He had also made many friends and become a star in the school football team.

But, just as they thought they had settled here, the family received a letter from the UK Immigration Service telling them that they were to be **deported** (forced to leave the country). It gave them a time and date to catch their plane from Heathrow Airport back to Beijing. Zhao took the letter to show his primary school headteacher and burst into tears. The head at Astley High also got to hear about it – that was when the two schools got involved.

The heads at Lyndhurst Primary and Astley High Schools promised that they would do their best to keep the children at school and prevent the government from deporting the family. Pupils collected 3,000 signatures on a petition that was sent to the government Home Office appealing for the family to be allowed to stay. At Astley High pupils wrote letters to their MPs and to the government to express their views. One girl, who was a friend of Miao, wrote

If the deportation takes place, it won't only be me who will feel upset and hatred towards the government but the entire school, her friends and their families. Remember these are real people with real feelings. This is real life. Please don't ignore it.

The campaign seemed to have an effect. The government delayed the deportation and at the time this book was written Zhao and his family were still living in Manchester. However, they still don't know if they will be allowed to stay permanently in this country.

Tanks Crush Student Protests in China

5th June 1989

A Flashback to June 1989 when a lone student stood in front of tanks in Tiananmen Square. You can find out more about China today in Unit 6 of this book.

The 1951 United Nations Convention on Refugees says that a person is a refugee if they are in genuine fear of persecution in their own country for reasons of:

* race
* religion
* nationality
* membership of a particular social group
* political opinion.

(Persecution can be a threat to life or freedom, or various forms of torture.)

When a refugee arrives in any country that has signed the UN Convention on Refugees they are entitled to seek **asylum** (a place of safety). While their claim is being examined they are known as **asylum seekers**. The government of that country cannot deport a genuine refugee back to their country of origin if their life or freedom would be in danger.

B Refugees and asylum seekers

Activities

1 Read the story of Zhao Hong Su and his family. Then look at source B.

   a) From the information that you have been given, should Zhao and his family be recognised as refugees and given asylum in the UK? Give reasons.

   b) What other arguments could you use against their deportation?

2 [L] *Either* Imagine that Zhao is one of your school friends. Write a letter to the government arguing that he should be allowed to stay in the UK. Use your ideas from activity 1 to make your letter as persuasive as possible. You can use the writing menu on page 19 to help you to write a persuasive letter.

   [ICT] *Or* On a computer, use the internet to find the 'Schools Against Deportations' website, set up by the Institute of Race Relations (IRR) at www.irr.org.uk/sad/. It gives up-to-date information about other children in the UK who are asylum seekers or refugees. You could write a real letter to the government on behalf of one of these children. If you do this, send a copy of your letter to IRR so that they will know you are supporting their campaign.

BUILDING BLOCKS

Can we believe everything we read in the papers?

You have read Zhao's story, so now you know a little bit about him and his family. Perhaps you have friends who are refugees. Refugees and asylum seekers are real people just like you. You might even be a refugee yourself!

However, newspapers often create myths about refugees and asylum seekers. We can forget that these are all real people. We read stories that lead us to believe that, somehow, refugees will ruin the country. It is not surprising that, after reading these stories, people can be suspicious, or even hostile, when refugees move into their area. The sources on these pages will help you to disprove some of the myths that you might read in the papers.

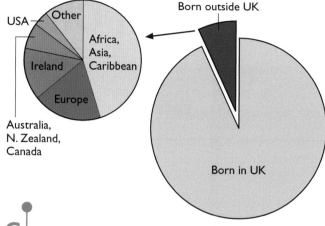

C Graphs showing where people currently living in the UK were born. Source: Office of National Statistics

Assignment

Here are some myths about refugees that you might have read in the newspapers.

THEY ONLY COME HERE TO SPONGE OFF US AND TO LIVE IN LUXURY

There is no control over who gets in to the UK

The whole country will soon be swamped by refugees

They don't do anything useful. We don't need them here.

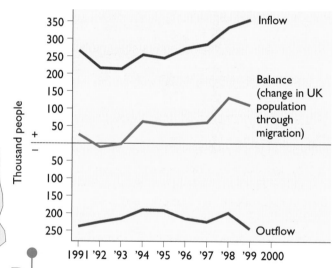

D Migration into and out of the UK since 1991
Source: Office of National Statistics

Work in a small group.

Study all the sources on these two pages. Look again at what you read on pages 8–11. Find evidence to disprove each myth. If you work in a group of four you could choose one myth each, or you could work together.

If you need any further evidence you could try some of these websites:

- Refugee Council at www.refugeecouncil.org.uk
- Commission for Racial Equality at www.cre.gov.uk
- Home Office Race Equality Unit at www.homeoffice.gov.uk/reu

You are going to use all the information to produce a leaflet called, 'Refugees – what can we believe?'

a) Divide your leaflet into four sections – one for each myth. Write the myths so that they stand out on the page.

b) Beside each myth draw or stick the evidence that disproves it. This could be a graph, a photo, a case study or any other evidence.

c) Write a paragraph to explain how the evidence helps to disprove each myth. Keep to the point. Leaflets must be easy to read.

ICT You could use a desktop publishing package on a computer to produce your leaflet.

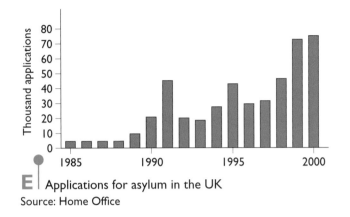

E | Applications for asylum in the UK
Source: Home Office

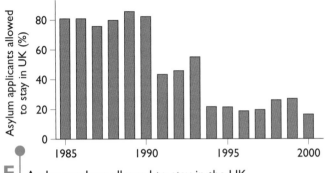

F | Asylum seekers allowed to stay in the UK
Source: Home Office

Support is provided to those asylum seekers who are destitute (have absolutely no money at all). Accommodation is provided in parts of the UK where there is no shortage of housing. Asylum seekers have no choice about where they live. They are given vouchers for food and other goods and £10 cash per week.

Asylum seekers can stay in the UK until the government considers their application to live here. If the application is refused they have the right to appeal. If the appeal fails they have to leave. The government aims to make each decision within two months and most appeals within a further four months.

G | Home Office guidelines on asylum seekers, 2001

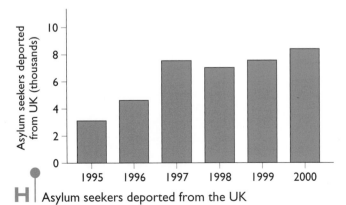

H | Asylum seekers deported from the UK
Source: Home Office

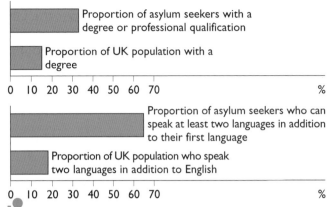

I | Comparison between asylum seekers and the rest of the UK.   Source: Refugee Council

J | Many of the essential services in the UK, such as the Health Service, have a serious shortage of workers. Doctors and nurses are recruited all around the world to work here. Many refugees have professional qualifications from their own countries.

K | Most asylum seekers are homeless when they arrive in the UK. The government houses them in temporary accommodation until it decides whether they will be allowed to stay.

BUILDING BLOCKS

In this Building Block you will think about the UK and what makes it a country. You will compare one part of the UK with the whole country and decide whether it should become independent.

1.6

The UK – one country or more?

How would you define what a country is? Easy, you might think. It's an area of land, with a border around it, that has its own government. And, of course, it should have its own flag (source A). But wait a minute! What about the United Kingdom? Is it one country, or is it really four separate countries? All four countries that make up the UK – England, Scotland, Wales and Northern Ireland – have a border around them and three of them have their own government. Only England does not have a separate government of its own.

You only need to ask a few people about their **nationality** to see how confused we all are. Some of us think of ourselves as British, while others would rather be called English, Scottish, Irish or Welsh. And what about the people that were not born here or don't feel they are any of these? It all adds up to a bit of an identity crisis!

A The Union Jack is the UK's national flag. It is really a combination of the flags of England, Scotland and Northern Ireland. It was also the flag that flew over the British Empire when Britain ruled over one third of the world's land surface.

B Who do we think we are? Four people living in the UK were asked about their nationality.

Hussain Mohammed

> I call myself British because I was born in England, grew up in Scotland and I like Scottish things. I drink Irn Bru and support Celtic.

Patrick McGuiness

> I'm Irish. I'm from County Down, my accent is strong and Irish and there's nothing about me that's British.

Paulina Dandgey

> I call myself Scottish because I was born in Glasgow and this is the country I'm most proud of. Britishness is fine but it's a bit too English.

Vanessa Ainscough

> I'm British – if you were born here you're British, if you weren't you're not. My dad was born in Jamaica – so what?

Language

How can we keep the English out?

Talk to them in Welsh.

Government

Hear, hear!

Religion

Natural resources

UK Fuels

Culture

Stand up if you love England!

Football team

Capital city

Natural boundary

Welcome to Wales
Croeso i Gymru

C Which of these should a country have?

So, now how would you define what a country is? Not quite so sure? Source C may give you some ideas about what the basic ingredients should be. You could use these ideas later in this Building Block when you have to decide whether or not one part of the UK should become independent.

D Welcome to Wales. Does the fact that Wales has its own language mean that it should be a separate country? So far, you don't need a passport to cross this border!

Activities

I Look at photo A.
  a) What images or ideas come to your mind when you see the Union Jack? Does it make you feel proud, or not? Why?
  b) Design a new flag for the UK for the twenty-first century. Try to represent everyone who lives here.

2 Look at source B.
  a) How would you describe your own nationality? Give reasons for your answer.
  b) How do you think any of the following differences may affect people's view of their nationality? In each case give reasons:
  i) young/old
  ii) white/non-white
  iii) born in England/born in another part of the UK.

3 Work with a partner. Look at source C.
  a) Decide which of these things a country needs in order to be a country. Rank them in order of importance. You could do this as a diamond rank exercise. The most important item is ranked 1, followed by the next two most important, ranked 2, with the least important ranked 5 at the bottom.
  b) Apply these ideas to the four countries in the UK and to the UK itself. Make a table with eight columns and five rows. The items in source C are the column headings. Put a tick or a cross in each of the boxes. For example, Wales has its own language, culture, football team, capital city and natural resources. It does not really have its own religion or a natural boundary. It has a government with some powers.

15

BUILDING BLOCKS

How do countries in the UK compare?

In the past twenty years people in different parts of the UK have become more aware of their nationality. In Wales there has been a revival of the Welsh language. There have been calls for **independence** in Scotland, and for Northern Ireland to be reunited with the rest of Ireland.

In 1997 the UK government responded to these pressures by agreeing to **devolution**. It handed some of its powers to new governments in Scotland, Wales and Northern Ireland. They each have control over things such as education and social services within their own border, but

London retains control of the most important things, such as defence and economic policy for the whole of the UK.

The UK government may hope that devolution will reduce the pressure for further changes. But some people believe that it will have the opposite effect. Now that people in Scotland, Wales and Northern Ireland have tasted power they are more likely to want full independence. It is possible that devolution could lead eventually to the break up of the UK.

	Population (millions)	Population change (% 1991–7)	Employment (% of total workforce)			Unemployed (%)	Wealth (average weekly income/£)
			Farming and mining	Manufacturing	Services		
England	49.3	+2.2	0.5	18.0	81.5	3.7	390
Northern Ireland	1.7	+4.5	3.0	18.0	79.0	4.3	302
Scotland	5.1	+0.3	1.9	16.4	81.7	4.5	384
Wales	2.9	+1.2	0.6	22.2	77.2	3.8	344
UK	59.0	+2.1	0.7	18.0	81.3	3.8	383

E England, Scotland, Wales and Northern Ireland compared
Source: Office for National Statistics

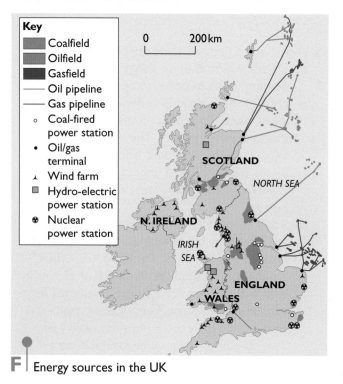

Key
- Coalfield
- Oilfield
- Gasfield
- Oil pipeline
- Gas pipeline
- ○ Coal-fired power station
- ● Oil/gas terminal
- � Wind farm
- ■ Hydro-electric power station
- ☢ Nuclear power station

0   200km

SCOTLAND

NORTH SEA

N. IRELAND

IRISH SEA

ENGLAND

WALES

F Energy sources in the UK

Activity

N Choose one country in the UK. It could be where you live. You are going to compare it with the UK as a whole.

a) Find your country in table E and locate it on map F.

b) Write a report comparing your country to the whole of the UK. Divide the report into six paragraphs about population, population change, employment, unemployment, wealth and energy resources. You can use the structure below to help you to write the first paragraph. This example is for Scotland.

The population of Scotland is 5.1 million, while the whole UK has 59 million people. This means that just under 10% of the UK's population lives in Scotland.

Or if you feel very confident you could write a report to compare all four countries in the UK. Which has the largest population? Which has the smallest population? Which has the fastest growing population? And so on.

The English think that Scotland couldn't stand up for itself in the big wide world. Of course, they are wrong! Scotland is big enough and rich enough to look after itself. If Scotland was an independent country it would be larger than the Netherlands, Belgium and Switzerland and it would be wealthier than Portugal, Greece or Ireland, to name just a few countries. Devolution is just a start. We want full independence!

Scotland can produce a lot more than whisky and haggis! It is a major energy producer. We export oil from the North Sea and we produce more than enough electricity to meet our own needs, especially from hydro and nuclear energy. We are strong in the financial sector too. Scotland has two large independent banks and many large insurance companies and investment groups. Edinburgh is one of the major financial centres, not just in the UK, but worldwide. It is also a beautiful city. So, who says that Scotland can't stand up for itself?

G | The Scottish nationalist view

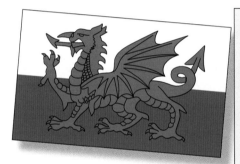

Wales has been ruled by England since 1277 but, at heart, the Welsh are an independent people. As recently as the 1960s it looked as if the Welsh language might die out. Now it is taught in schools and many pupils do all their subjects in Welsh. More people speak Welsh now than 40 years ago.

The Welsh people love their country. For much of the twentieth century its beauty was sacrificed to meet the needs of the British economy. The landscape was ravaged by coal tips and slag heaps (from steelworks), drowned by reservoirs, smothered by coniferous forest, shelled by army tanks and polluted by nuclear fall-out. All in the name of progress, but it was mainly progress for people living in English cities. Now we want to reclaim our landscape as well as our language.

Wales sees its future in Europe, with or without the rest of the UK. The European Parliament is a place where small nations like Wales can get together to make their voice heard.

H | The Welsh nationalist view

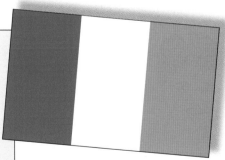

You can't go against the tide of history forever. The British ruled Ireland for centuries but eventually, in 1922, they had to give it independence. Unfortunately, as part of the deal, the six counties in the north had to remain part of the UK to keep the Protestant majority happy. That is how Northern Ireland was formed. The new government made laws that were unfair to the Catholic minority and that was when the troubles began. Since 1969 the British Army has been in Northern Ireland and, for most of the time, we have been governed directly from London.

Now, there is a sort of peace and the warring factions have put away their guns. But everyone knows that it could all come to an end. The Catholic population is growing faster and will soon outnumber the Protestants. Pressure for change on the British government will grow. The tide of history is strong and there won't be real peace until we are part of a united Ireland again.

I | The Irish nationalist view

Should we go independent?

Nationalism is the love of your own country. You can read the views of three nationalists in sources G, H and I. But it is not just the Scottish, Welsh and Irish that can be nationalists – the English can too (photo J)! Throughout Europe, as well, there has been a growth of nationalist feelings. Since 1989 fourteen new countries have appeared on the map of Europe.

So, what has led to this trend? Often nationalism grows in remote regions, far from the centre of national power, where people feel they are ignored by their government. In Europe these regions feel they have a better chance of being recognised by the European Parliament than they do by their own governments. And, in a world that is growing ever more complex, people look to their culture and traditions to give them a sense of identity.

J English football supporters waving the English flag at Euro 2000.

Activity

L You are going to analyse the views of the three nationalists in sources G, H and I.

a) Identify the facts and the opinions in each one. (A fact is a statement that can be proved, an opinion cannot be proved.) On a copy of the three views underline the facts and the opinions. Use two colours. You can make use of the facts in the assignment below.

b) Identify the techniques that the three writers use to persuade us. Find examples of words and phrases where the writers use:

- positive language – words like *strong* or *major*, to make their ideas sound important
- emotive language – words like *unfair* or *beautiful* to get the reader on their side
- humour – to make the reader smile and be more likely to listen to their ideas
- specific examples that support their argument.

You can use the same techniques in your assignment.

Assignment

I believe that Scotland should be independent ...

You are a politician representing one part of the UK. Together with the rest of your class you are going to debate whether or not your country should become independent from the rest of the UK. (Use the same country that you chose for the activity on page 16.)

First, you must decide what your own opinion is. Are you for independence or against? Think about what you have learnt in this Building Block.

In activity 3 on page 15 you considered what things a country needed in order to be a separate country. Does this make the case for or against your country's independence?

On page 16 you compared your country to the whole of the UK. Would it be able to stand on its own?

Finally, on page 17 you read the views of three nationalists. How far do you agree with their opinions? Which facts are important?

Write a short speech (to last about two minutes) that you can make in the debate. The writing menu on the next page will help you.

Your teacher will chair the debate to make sure it is conducted fairly. At the end they may remind you of the best points on both sides of the argument and ask you to vote for or against independence.

Scotland could never be independent ...

How to write a speech for a debate

Text type	Tense	Starters	Links	Conclusions	Vocabulary
Recount	Past	First, …	… because …	In conclusion, …	country
Description	*Present*	Second, …	… that shows …	In summary, …	border
Method	Future	In the first place, …	… therefore …	Overall, …	government
Explanation		Importantly, …	… in fact …	On the whole, …	nationality
Persuasion		For example, …	… it is clear that …	In short, …	language
Discussion		Obviously, …	… not only … , but also …	In brief, …	culture
		Finally, …	… furthermore …		resources
					independence
					devolution
					nationalism

STRUCTURE FOR WRITING A SPEECH

1   When you write a speech for a debate remember that you must try to *persuade* people. Start by stating what your point of view is. For example, *I believe that Scotland should be independent.*

2   Write your speech in the way that you would talk. Keep your sentences short and don't go into unnecessary detail. Remember how boring it can be if your teacher talks for too long! Read your speech aloud to hear what it sounds like. Don't rush it. Slow down when you get to the important bits. Check that it is the right length.

3   There are techniques that you can use to make your arguments more persuasive:
   • Use positive language rather than tentative language. For example, *An independent Scotland will clearly be . . .* rather than, *If Scotland becomes independent it may be . . .*
   • Use emotive language to get the listener onto your side. For example, *How can it be fair for Scotland's future to be decided 600 km away in London?*
   • Humour will help to get your audience's attention. For example, *Have you heard the one about the Scotsman and the Englishman?*
   • Finally, if you really want to convince your listeners, give specific examples to support your arguments. For example, *Scotland doesn't need to depend on England. It has enough oil in the North Sea to last a lifetime.*

4   Conclude your speech by reiterating what you said at the start. Summarise the main arguments that you have made during the speech to help your audience to remember.

1.7 Do we want to be part of Europe?

Why is it that the most boring things in life so often turn out to be the most important? I need only mention maths to prove my point!

Take Europe as an example. The **European Union** (EU) now has its own currency: the **euro**. The UK's government has to decide whether or not we join. It could be the most important decision that we ever have to make. It will affect prices in the shops, wages, the sorts of job we do, our lifestyles, where we go on holiday. Above all, it will affect whether we are British or fully European. So, if it is that important, why is it that whenever a politician mentions Europe on TV, most of us switch off? Probably, for the same reason that you fall asleep in maths!

The EU (it was called the European Community in those days) was formed in 1957 by six countries to create a market for the goods they produced. The UK joined in 1973. Since then we have seen faster economic growth and enjoyed greater prosperity than ever before. There are now fifteen countries in the EU and it is likely to expand further in the next few years as countries from Eastern Europe join (map B).

A Should we take the plunge?

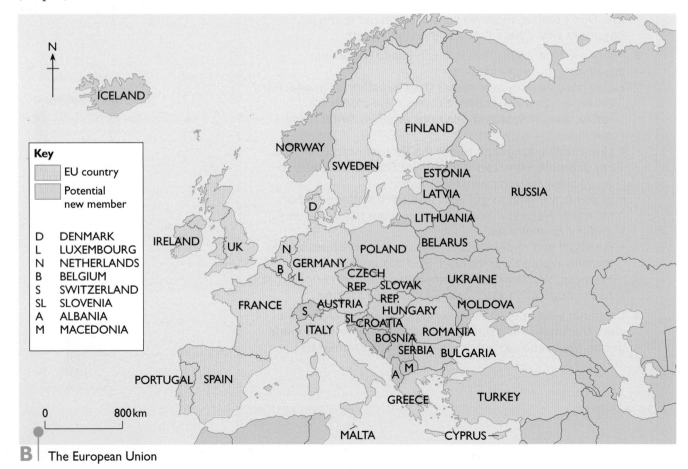

B The European Union

Key
- EU country
- Potential new member

D   DENMARK
L   LUXEMBOURG
N   NETHERLANDS
B   BELGIUM
S   SWITZERLAND
SL SLOVENIA
A   ALBANIA
M   MACEDONIA

Arguments in favour of integration

- Will help the UK to trade more freely with other EU countries, giving producers a larger market for their goods
- More trade will lead to a stronger economy and greater prosperity
- The UK will have to conform to EU laws on living and working conditions leading to improvements for those who suffer poor conditions
- There will be more co-operation with other EU countries on issues like drugs, refugees, terrorism and environmental pollution
- There is more likely to be peace in Europe if all countries co-operate than if they all pursue their own interests

Arguments against integration

- The UK government will lose some of its powers and the UK will also lose its own currency (the pound)
- The EU is less democratic than the UK, since some of its most important figures are not elected
- The UK will lose part of its social and cultural identity because the EU will impose its own laws and standards
- Better working conditions imposed by the EU will make UK industry less competitive compared to other parts of the world and jobs will be lost
- EU expansion to include poorer Eastern European countries will drain money from relatively wealthy countries like the UK

C Arguments for and against integration of the UK in Europe

D Stages in the growth of the European Union

Countries in the EU are getting closer economically and politically. This process is known as **integration**. The question of integration with Europe divides the main political parties in the UK. Some feel that unless we join the Euro and get more involved in Europe we will be left behind. Others feel that integration would mean that the UK would lose its ability to govern itself. The main points in the debate are outlined in table C.

Activities

1 Look at map B and graph D.
  a) Make your own map showing the growth of the European Union. It should include the information on graph D. Shade the map using different colours to show the countries that joined at each date. Also show the countries that are likely to join in future.
  b) Describe the pattern shown on your map. How has the European Union grown?

2 Work with a partner. Look at table C. Discuss the ideas in the table and decide whether or not you are in favour of closer integration with Europe. One way to do this would be to cut out copies of the ideas in the table and rank them in order of their importance for you, from 1 for the most important to 10 for the least important. Place the numbered statements back in their original column to see which side has the most important ideas. Are you for closer integration, or against?

DIGGING DEEPER

Why is Europe so important?

The only way this company will survive is if we can sell more of our cars in Europe.

E | Midlands car factory worker

Times are hard for farmers. They will get even harder when the EU expands and there is more competition.

G | Welsh hill farmer

If the UK government does not treat me fairly I can appeal to the EU. As an EU member the UK has to listen to what Europe says.

F | Asylum seeker recently arrived in the UK

The only way I can make this job pay is by working longer hours. But that is illegal in Europe.

H | Long-distance lorry driver

Activities

1 Look at photos E–H.
   a) Suggest whether each person would be for or against integration with Europe.
   b) Look back at table C on page 21. Which might be the most important idea that each person would use to support their opinion?

Homework

2 Carry out a survey to find out how people you know feel about UK membership of the EU. You will need to prepare a survey sheet. Interview a range of people you know, including people in different age groups. Try to interview at least ten people. Ask them to answer these two questions:
   a) Do you think that the UK's membership of the EU is an important issue? Yes or no?
   b) Do you think that the UK should:
      i) have closer integration within the EU
      ii) continue to be an EU member without closer integration
      iii) withdraw from the EU?
Ask them to give reasons for their opinions.
Share your results with the class. Based on your survey, do people think that Europe is an important issue? Are people in the UK in favour of integration with Europe, or not? What are the most common reasons that people give?

Assignment

The UK government wants to hold a referendum before it agrees to closer integration with Europe and joining the euro. (In a referendum the government refers an issue to the people to decide in a special vote.) But will people bother to vote? The result of your survey should give you a good idea about people's attitudes.

Produce an advert that the government could use before a referendum on Europe to persuade people that it is important to vote. It could be an advert to go on a roadside billboard, a full-page newspaper advert or a television advert.

TRANSPORT ISSUES –
Are cars the solution or the problem?

These days, most people in the UK travel by car. The number of road journeys we make has doubled over the past 25 years. But, rather than getting quicker, our journeys are actually getting slower.

- Have you ever been stuck in a traffic jam as bad as this one? Where was it?
- How did it make you feel?
- What problems does all this traffic create for people? And for the environment?
- What are the solutions, do you think?

2.1 How did you get to school today?

Did you get up late this morning and ask your mum or dad for a lift to school in the car? Maybe you travel to school every day in a car because you have far to go, or there are no buses or trains near your house, or even (I'm sure that this is not true of you!) because you don't like walking. If so, you are one of a growing number of pupils who travel to school by car (see graph B).

The government is worried about the number of school journeys made by car. This is what it says:

'Not walking or cycling to school means that children get much less exercise and leads to car dependency at an early stage in a child's development. These children will find it harder as adults to use cars responsibly and will have fewer opportunities to develop the road sense they need as pedestrians or cyclists.'

Source: Department for Environment, Transport and Regions, 1998

Do you think the government is right to be so worried?

A 'Bye dad! Thanks for the lift'

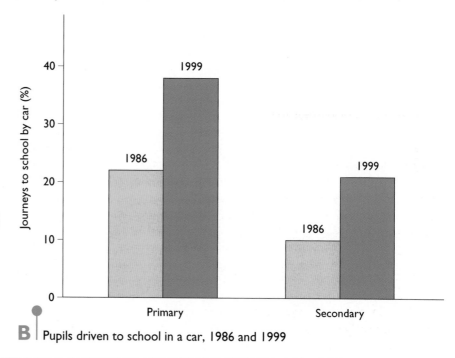

B Pupils driven to school in a car, 1986 and 1999

Activities

1 Work with a partner.
   a) Think of all the advantages of using a car to get to school. Make a list of your ideas. Then make a list of the disadvantages (don't forget to think about how cars can affect other people and the environment).
   b) Draw a table like the one opposite to show your ideas. One advantage and one disadvantage have been done for you.

Advantages of going to school by car	Disadvantages of going to school by car
It is usually quicker (if there are no traffic jams!)	It adds to traffic congestion

2 Look again at the table you have produced. Do you think the government is right to be concerned about the number of pupils being driven to school? Give at least three reasons for your answer.

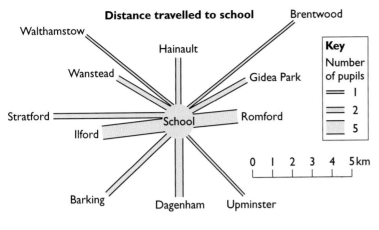

Distance travelled to school

Method of transport

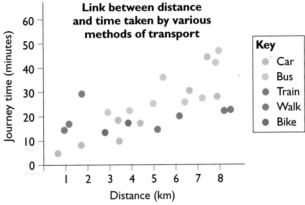

Link between distance and time taken by various methods of transport

C Class survey of journeys to school

A group of Year 7 pupils at one school carried out a class survey to investigate journeys to their school. They wanted to find out how pupils got to school, how far they travelled, and if there was any link between the distance they travelled and the type of transport they used. Source C shows their results.

3 Study source C.
  a) Write down what questions pupils would have asked to obtain this information.
  b) Is there any other information that you would want to find out from a class survey about journeys to school? What other questions would you need to ask?

4 Carry out your own class survey about journeys to school.
  a) Design a survey sheet to do your class survey. Use the questions you thought of in activity 3. Try to design a single sheet that you could use to survey all the members of your class.
  b) Use your survey sheet to ask members of your class about their journeys to school.

5 Display the results of your survey. You could use the same types of graph as the ones in source C, or you could choose your own types of graph.

6 *Either* write conclusions about the results of your own class survey, *or* write conclusions about the survey results in source C. First, write a paragraph to describe how pupils got to school.
   You can use this writing frame.

The majority of pupils in the class got to school by _____. _____% of the pupils use _____. _____ % came by bus or train. Only _____% of pupils _____ to school. Even fewer pupils _____.

Now write two more paragraphs to answer the following questions:

• How far do pupils travel to school?
• What link is there between the distance travelled and the type of transport pupils use?

2.2 On the move in the UK

It is almost impossible to imagine the changes in transport that have happened in this country over the past 50 years. If your grandparents were in the UK they may remember trips to the shops by tram and long journeys by steam train. Owning a car 50 years ago was quite unusual. The first **motorway** in the UK – the M1 – was opened in 1961. Since then the motorway **network** has spread its tentacles across most of the country (map A).

At the same time, there has been a massive increase in car ownership. Nowadays it is common for a family to own two cars. All over the UK we have problems of traffic **congestion**. Even motorways cannot cope with the volume of traffic.

Now the government wants to persuade people to switch back to using **public transport** (buses and trains). In cities, new forms of public transport are being used to encourage people out of their cars. Even trams are making a comeback!

Activities

1  Look at map A.

   a) Plan the easiest route from your nearest town or city to London by road. If you live in London, choose another city to go to.

   b) Measure the distance that you would have to travel. You can do this using a strip of paper or some string. First, bend the paper or string to follow the route that you would take. Mark the distance on the paper/string. Then straighten it out and place it on the scale line to find out the real distance.

   c) Work out how long the journey would take if you travelled at an average speed of 80 km per hour.

A The UK motorway network

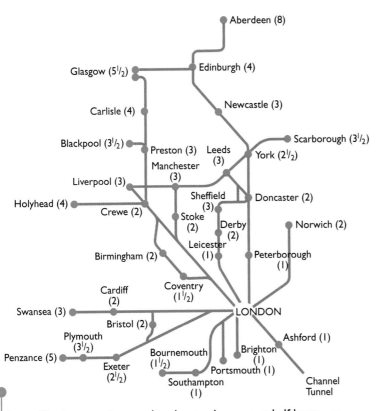

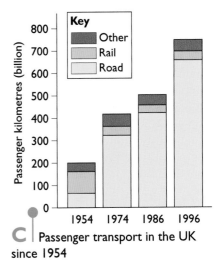

C Passenger transport in the UK since 1954

B Inter-City journey times to London to the nearest half hour

D The Eurostar train will soon be able to do the journey from London to Paris in two hours.

2 Look at map B.
  a) How long is the journey time by rail from your nearest city to London? How does this compare with the time by road?
  b) Choose five more cities and find the journey time by rail to London. Then work out the time by road (using map A and assuming an average speed of 80 km/hour). Compare the times.

3 Look at the box below. It gives some advantages and disadvantages of travelling by road and rail.

> traffic jams   relaxing journey   slower
> travel from door to door   road accidents
> faster   few rail accidents   travel any time
> travel at fixed times   low cost per journey
> high cost per journey   more pollution
> less pollution   timetable delays
> tiring journey

  a) Sort them into four groups and list them in a table like the one below.

	Advantages	**Disadvantages**
Road		
Rail		

  b) Which of these people would be more likely to travel by road or rail? In each case give at least one reason.

  - A businesswoman going to a meeting in London
  - A sales rep travelling to meet clients around the Midlands
  - A commuter travelling to work each day in Manchester
  - A farmer going to an agricultural show in mid-Wales.

  c) Design an advert that the government could use to persuade people to switch from cars to trains.

4 Look at graph C.
  a) Describe the changes in transport shown by the graph.
  b) Explain why these changes have happened.
  c) Suggest what the graph will look like by the year 2020. Give reasons.

2.3 Gridlock in the city

Cities and cars don't work well together. Too many people and buildings get in the way – not to mention other cars! Traffic in London moves, on average, at less than 20 kilometres per hour – about the speed of a bicycle. As a result, most city centres are not easily **accessible** by car.

An alien that arrived from outer space and saw the traffic problems in our cities would probably conclude that we are mad! This is obviously the impression that the Brainians in source A have got.

The problem was one of transport.

The Brainians could see the long, thin arteries along which the humans travelled. They noted that after sunrise the humans all travelled one way and at sunset they all travelled the other. They could see that progress was slow and congested along these arteries, that there were endless blockages, queues, bottlenecks and delays causing untold frustration and inefficiency. All this they could see quite clearly.

What was not clear to them was why.

They knew that humanity was stupid, they had only to look at the week's top ten grossing movies to work that out, but this was beyond reason. If, as was obvious, space was so restricted, why was it that each single member of this strange lifeform insisted on occupying perhaps 50 times its own ground surface area for the entire time it was in motion – or not in motion, as was normally the case?

A | Extract from *Gridlock*, a novel by Ben Elton. (**Gridlock** is the point at which congestion becomes so bad that traffic comes to a complete standstill.)

Exhaust fumes from vehicles are the main cause of **pollution** in cities. This has been linked to health problems such as asthma.

People without cars – the young, the elderly and those who don't drive – become more isolated.

Road accidents kill hundreds of people in the UK each year. Most accidents occur in cities.

The amount of traffic is one of the main reasons that people give for wanting to leave cities. Could cars eventually lead to the death of our cities?

Commuters – who travel to and from work every day – waste hours sitting in traffic jams.

The huge volume of private cars and lorries causes delays for public transport and emergency services.

Car parking takes up huge areas of space in cities that are already overcrowded. More than one million cars come into the centre of London every day.

Noise, dirt and pollution from vehicles have turned city streets into no-go areas for people. They are unsafe places for children to play.

B | Transport problems in a busy city

You won't catch me sitting in a traffic jam. I can get to work in half the time it takes by car. I wear a helmet because cyclists are more likely to be injured in an accident than car drivers. And you can guess why I'm wearing this mask.

C

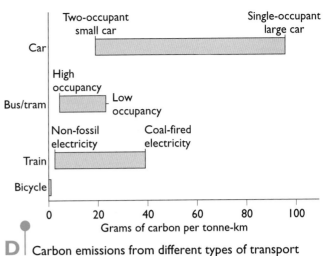

D | Carbon emissions from different types of transport

E | Car adverts are everywhere.

Activities

1 L Read extract A.

Work with a partner. Role play an imaginary conversation between a Brainian and a human being.

The Brainian could ask, 'Why do you use cars?' How would the human reply?

In return, the human might ask, 'Well, if you're so brainy, what would you do?' What answer could the Brainian give?

Think of more questions of your own. Use your ideas to write a script in your book for the conversation.

2 Look at photo B. Read all the information in the boxes.

Think about how each of these people in the city would feel about the transport problems. (Photo C gives you some ideas for a cyclist.)

| a car driver   a pedestrian   a bus passenger |
| a resident   a shop owner   a cyclist |

Draw a large spider diagram, like the one below. In each of the speech bubbles write how one of the six people would feel about transport problems. Give each one a title.

Transport problems

3 Look at graph D.
  a) For each form of transport, work out the average level of emissions (halfway between the high figure and low figure). List them in order, starting with the highest.
  b) Explain the difference between the high figure and low figure for each form of transport.
  c) Suggest where the emissions for cycling come from.

4 Look at photo E.
  a) How do adverts like this persuade people to buy cars?
  b) How does the advert differ from the reality of driving in a city?
  c) Design your own advert. *Either* design an advert to persuade people **not** to buy a car *or* an advert to persuade people to buy a bike instead of a car.

2.4 Carving up the countryside

More and more of the countryside is being covered with tarmac to make space for the growing number of vehicles. Many towns in the UK now have a **by-pass** to replace an old road that went through the town itself. Others have a **ring road** to relieve the pressure at the town centre. The M25 – one of the busiest motorways in the country – is a giant ring road around London (see map A on page 26).

Car ownership in the UK is predicted to double to 50 million cars by the year 2025. Parked nose to tail that is the equivalent of a 500-lane motorway from London to Edinburgh!

Over the past ten years there have been many protests against new roads. One of the biggest was at Twyford Down, an area of chalk downland on the route of the M3 motorway.

Most of the M3 was complete by 1993, apart from a short gap near Winchester. This led to long delays in both directions on the motorway. The government came up with three alternative routes to close the gap (see map A). There was a long **public enquiry**. In the end, the government chose the route through Twyford Down (photo C). Would you have made the same decision?

Activities

1 Look at map A. It shows three proposed routes for the M3 near Winchester. You have to choose the best route.
   a) On a copy of the table below, tick boxes to show the advantages of each route. For example, route C is the shortest route so you tick box C on that row. You may want to tick more than one box in some rows.
   b) Which route has the most ticks? Is this the route you would choose?
   c) Write a paragraph to explain your choice.

	Route A	Route B	Route C
Cost	£36 million	£47 million	£128 million
The shortest route			
Follows original road			
Less noise in Winchester			
Does not involve tunnelling			
Does not involve excavating			
Avoids River Itchen SSSI			
Avoids St Catherine's Hill SSSI			
Avoids Twyford Down AONB			
Total			

2 The government chose route A. You can see the environmental impact in photo C.
   a) Suggest why the government chose this route.
   b) Imagine that you are an environmentalist. Write a letter to the government to say what you think of their decision.

3 Environmentalists were angry at the government's decison. As a result, protesters like Swampy (in photo B) took direct action. Sometimes this involved breaking the law.

   In a small group, discuss these questions.
   a) Why do you think protesters took direct action?
   b) Were they justified in breaking the law? Why?
   c) Was the action effective? Why? (Think about the effect on public opinion.)
   d) Are there any issues that you would feel strongly enough to protest about? How would you protest?

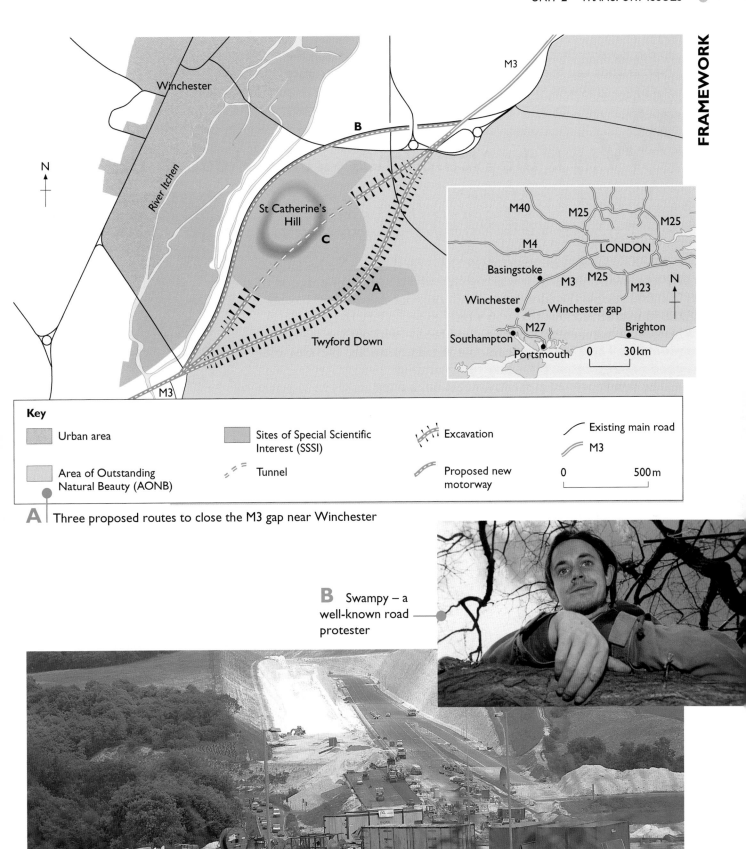

Key

- Urban area
- Area of Outstanding Natural Beauty (AONB)
- Sites of Special Scientific Interest (SSSI)
- = = = Tunnel
- Excavation
- Proposed new motorway
- Existing main road
- M3

0   500m

A Three proposed routes to close the M3 gap near Winchester

B Swampy – a well-known road protester

C Construction of the new section of the M3 at Twyford Down (route A on the map). The road opened in 1997.

In this Building Block you will compare urban transport in the Netherlands and the UK. You will use the ideas to investigate transport problems in your local area.

2.5 Why is the Netherlands streets ahead of us?

The Eldridge family from London have decided to do something different for their holiday this year – they have left their car at home! They have gone to the Netherlands – a country where it is much easier to travel without a car.

They are staying in Amsterdam, the capital of the Netherlands. By European standards it is a fairly small city with a population of 700,000 people (London has six million), but Amsterdam has all the tourist attractions that you would expect a city to have – history, culture, entertainment, good food – without the transport problems. There are fewer cars, but thanks to canals, bicycles and good public transport it is easy to get around.

The family arrived at Centraal Station, having flown from London City Airport to Schipol Airport near Amsterdam (see map B). They hired bicycles and took a short ride to their hotel. Now it is time to explore the city.

A A family cycling in the Netherlands

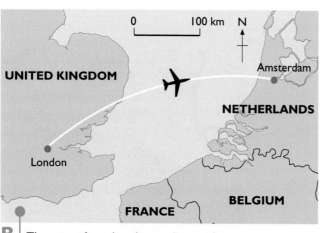

B The route from London to Amsterdam

Key

ℹ️	information	🌸	market
✉️	place of interest	▦	pedestrian area
🏛️	museum/gallery	📖	library
◉	landmark	🎵	music
𝕀	monument	✉	post office
✝☦	church/synagogue	⛴	museumboat
○	hotel	⛴	canalbus
🛡️	theater	🚆	train
⬕	shopping	🚋	tram

C Tourist map of central Amsterdam

Activities

1 Before going on holiday the Eldridge family received directions to get to their hotel from Centraal Station. Read the directions and follow them on the map.

> Turn left out of the station and cross the bridge to St Nicholaas Kerk. Cross the road to the church and turn left over the first canal bridge, which will bring you to Gelderskade canal. Turn right at Schreierstoren Hotel and cycle south for 500 metres keeping the canal on your right. You will pass Waag, a medieval gatehouse. Continue south for another kilometre beside the canal. The road leads to a bridge over the Amstel canal. Cross the bridge, turn right and follow the tram line across Muntplein to Reguliersbreestraat. Take care to avoid pedestrians as you cycle across Rembrandtplein to Amstelstraat! You will find the hotel on the left.

Mark the route on a copy of the map. What is the name of their hotel?

2 Work with a partner.

a) Each choose at least one place to visit in Amsterdam. It could be a museum, theatre, market, or whatever you want. Work out the best route from the hotel where the Eldridge family are staying. Mark it on your map.

b) Describe your route to your partner so that they can follow it on their map. They have to work out which place you have chosen to visit. Then listen carefully to your partner and follow their route. Can you work out what they chose to visit?

3 Look at map C and photo D.

a) What evidence can you find to show that it is easier to get around Amsterdam than many UK cities? Make a list.

b) What advantages for transport does Amsterdam have? Again give evidence from the map and photo.

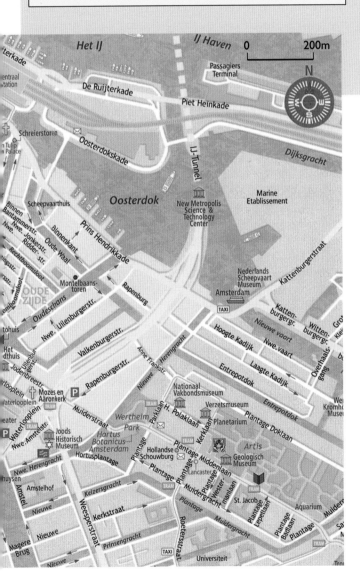

D Jordaan – an area close to the centre of Amsterdam.

What are the solutions to traffic congestion?

E Trams are the quickest way to get around Amsterdam. They run on overground rails in the centre of the road and don't get held up by traffic. It is possible to travel from almost any part of the city into the centre in less than 30 minutes.

F **Pedestrianised areas** are closed to motor vehicles and can make shopping a more relaxed experience. The main danger for pedestrians in Amsterdam is fast-moving bicycles!

G **Traffic calming** in residential areas slows down cars by using speed bumps, and pedestrians and cyclists are given priority at junctions. Speed limits of 15 kilometres an hour make it safer for children to play.

Throughout Europe the volume of traffic in cities has led to similar problems: road congestion, pollution and poor quality of life. The Netherlands was one of the first countries to look for solutions back in the 1970s. Their approach to solving the problems was very different to the way that we have tried to solve them in the UK.

In the Netherlands the priority was to reduce the number of vehicles in the city. They did this by improving the public transport system (photo E), and by making it harder to use cars in the city. Car parking became more expensive, pedestrianised areas (photo F) and cycle lanes were closed to cars, and strict speed limits were imposed (photo G).

In the UK the priority was to make it easier for people to drive their cars in cities. New roads were built to divert vehicles from congested areas (photo I), more car parking spaces were provided (photo J) and traffic management schemes were introduced to speed up the flow of traffic (photo K). But transport policy in the UK is changing. The government has realised that there is a limit to the number of vehicles a city can take. It has begun to follow the Dutch example and now wants to reduce the amount of traffic in our cities.

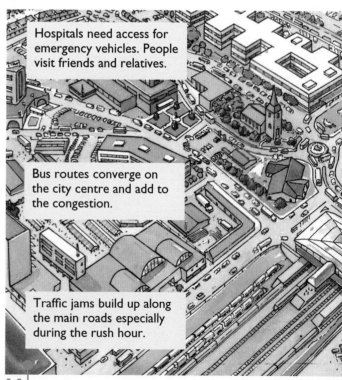

Hospitals need access for emergency vehicles. People visit friends and relatives.

Bus routes converge on the city centre and add to the congestion.

Traffic jams build up along the main roads especially during the rush hour.

H A congested city centre

Activities

Work with a partner. One person looks at photos E–G on page 34, which show measures to deal with urban transport problems in the Netherlands. The other person looks at photos I–K on this page, which show measures to deal with urban transport problems in the UK.

1 Look at drawing H and your photos. How would you each solve these transport problems?

Draw a map of the area in the drawing. Put your ideas for solving transport problems onto your map. Use different colours so that your ideas will show clearly. For example, if you want to build a ring road, draw the route on your map. Shade the buildings you would need to knock down and the roads you would close.

Label your map to explain the changes that you have made. Write at least one sentence to explain each change. Give your map a title.

2 Compare the map that you drew in activity 1 with the one your partner drew. If possible, stick a photocopy of their map in your book on the page opposite your own.
   a) Describe the differences between the maps.
   b) Explain the differences from what you know about transport priorities in the Netherlands and the UK.

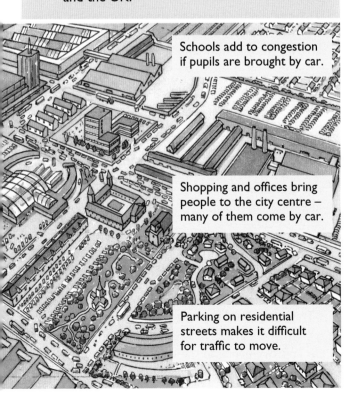

Schools add to congestion if pupils are brought by car.

Shopping and offices bring people to the city centre – many of them come by car.

Parking on residential streets makes it difficult for traffic to move.

I Urban ring roads divert traffic away from busy city centres. But they make it harder for pedestrians and cyclists to get to the centre. Underground subways can be dark and dangerous places.

J Multi-storey car parks take up space in the city centre. They encourage people to drive into the city and create more congestion. The buildings are usually rather ugly.

K **Traffic management** schemes, such as roundabouts and one-way systems, help to speed up the flow of traffic on narrow streets. But they can be dangerous for pedestrians and people who live there.

BUILDING BLOCKS

What local transport problems do you have?

What are the transport problems in your local area? Are the buses always late? Is it too dangerous to cycle because the cars go too fast? Or maybe the cars don't move at all! Whatever the problem, here is an opportunity for you to carry out an investigation. And, you don't need to live in a city. You can investigate transport in rural areas too.

When you have written up your investigation you could send a copy to your local council, together with your ideas for improvements, to see if they will do anything to sort the problem out.

L Pupils recording traffic flow outside their school

Assignment

You are going to investigate a transport problem in your local area. You could use the results of the investigation to recommend how the problem should be solved.

1 Work in a group of about four pupils.

Identify a transport problem you want to investigate. The easiest problem to investigate will be one that is close to your school. This will allow you to carry out the investigation at different times of the day, or days of the week.

Think of a question that your investigation could focus on. For example:

How much traffic is there on Earlham Grove?
What are the busiest times of day on Earlham Grove?
Which parts of Earlham Grove get most congested?

You may decide to investigate more than one question. Whatever you decide, check with your teacher to find out whether it will be possible to do.

2 Plan how you will carry out the investigation. For example, to investigate the question *How much traffic is there on Earlham Grove?*, you could do a traffic count.

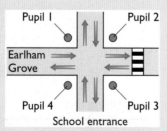

You will probably need one person to stand on each side of the road, or each corner of the junction, to record the flow in one direction each. Each person would need a sheet on a clipboard and a stopwatch to record the traffic. Would you want to record cars, lorries, buses and other types of

vehicle separately? Do you need to record the traffic flow at different times of day?

Another part of the investigation could be to find out people's opinions. You could interview different groups of people such as pedestrians, cyclists, residents and car drivers. You will need to prepare questions for your interviews and sheets on which you can record the answers.

Check your plans with your teacher, then carry out the investigation. Remember, **roads are dangerous**. It is best to have adult supervision when you do your investigation.

3 Think of the best ways to show your results. One way to show traffic flow is to use proportional arrows whose width indicates the number of vehicles. This is how you would show the traffic flow at the junction on Earlham Grove. You might draw more diagrams like this to show traffic flow at different times of day. You could show the results of your interviews in the form of tables or graphs.

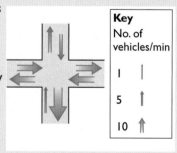

4 Write up your investigation. You can use the ideas in the writing menu on page 37. You could use a word-processing package on a computer to make it look professional. Don't forget that you could send a copy of the investigation, together with your ideas for improvements, to the local council. So make it good!

How to write up your investigation

Text type	Tense	Starters	Links	Conclusions	Vocabulary
Recount	*Past*	First, …	… and …	In conclusion, …	traffic
Description	Present	Second, …	… also …	In summary, …	volume
Method	Future	Next …	… too	Overall, …	flow
Explanation		Before …	… as well as …	It appears …	route
Persuasion		While …	… and then …	It is clear …	access
Discussion		Later …	… in addition …	It is recommended …	congestion
		Finally, …	… so …		vehicles
			… because …		rush hour
			… due to …		solution
			… therefore …		traffic calming

STRUCTURE FOR WRITING AN INVESTIGATION

1   Title and aims
Start with the question that you set out to investigate. For example, *How much traffic is there in Earlham Grove?* Explain the aims of your investigation. For example, *to identify local traffic problems and how they affect different groups of people.*

2   Method
Write the method in the past tense. Normally you write your method in the first person plural – *we identified the question, we planned our investigation.*
Start with your preparation before the investigation – *Before we did the investigation* . . . What did you do in the classroom? What equipment did you need? What recording sheets did you have to prepare?
Then describe the things you did in the investigation in chronological order (the order in which you did them). *First, we split up and one member of the group stood on each corner* . . . Include a map to show where you did your investigation.

3   Results
Present your results in the form of maps, tables or graphs. Give each one a title and, if necessary, a reference – *Figure 1, Table 2.* When you write about your results you can use these references to illustrate what you mean – *Figure 1 shows that the traffic flow* . . .
When you write about your results change from the first person to the third person – *The traffic was busiest* . . . rather than *We saw* . . . You are now writing about the **results**, not about what **you did**. Write a paragraph about each table, map or graph you have drawn.

4   Conclusions/recommendations
Before you start this, go back to the question in your title to remind yourself what the investigation was about – e.g. *How much traffic is there in Earlham Grove?* This should be the question that you try to answer in your conclusions. Use the results of your investigation to help you to answer it.
At the very end, make your recommendations about how to solve the transport problem you set out to investigate.

GROUNDWORK

FRAMEWORK

BUILDING BLOCKS

DIGGING DEEPER

BUILDING BLOCKS

In this Building Block you will examine the need for a new motorway to relieve traffic congestion around Birmingham and produce a news report about the issues.

2.6

Does Birmingham need another road around it?

... Five Live Travel News ...

Motorists heading towards Birmingham should expect severe delays on many of the motorway approaches. It's all down to sheer weight of traffic, with people rushing out to do last minute Christmas shopping. As usual the M6 between junctions 4 and 12 is badly affected in both directions. There are tailbacks on the M42 north and south of the M6. The M5 has a five-mile queue northbound heading towards the junction with the M6. All roads leading to junction 6 – that's Spaghetti Junction – are heavily congested. Avoid the whole area around the Merry Hill Shopping Centre near Dudley. If you're heading off for the Christmas holiday, we advise you to plan your route carefully. And if your journey isn't really necessary, why not stay at home and stuff your turkey!

Joanne Sale – Five Live Travel

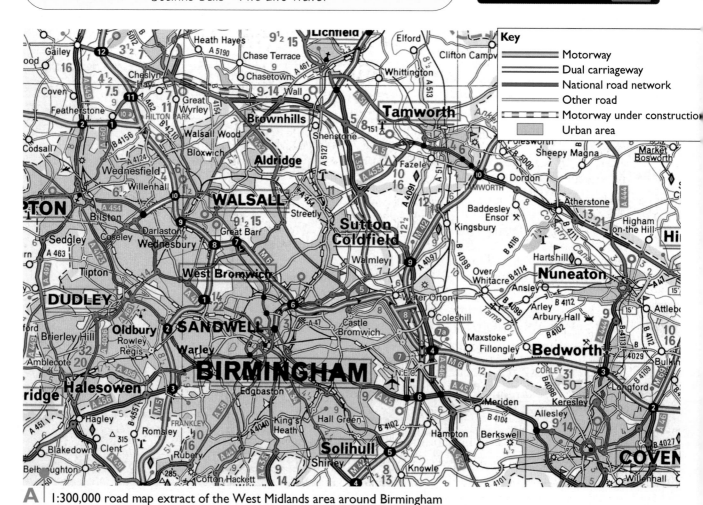

A | 1:300,000 road map extract of the West Midlands area around Birmingham

The West Midlands lies at the heart of the motorway network (see map A on page 26). It has some of the busiest roads in the country. Long distance traffic from all corners of the UK converges on the West Midlands, where local traffic also uses the motorway to avoid congestion in Birmingham. The main problem is the large number of intersections on the M6 that allow local traffic to slip on and off the motorway. At busy times (which can be at almost any time) this causes long traffic jams and untold frustration for drivers (see photo C).

For many years there have been plans to build a new road to relieve the congestion on the overcrowded section of the M6 around Birmingham. The government held a public enquiry to find out opinions. On the one hand, businesses argued that a new road was essential if they were to survive. On the other hand, environmentalists argued that a new road would create more traffic and take up more green land around Birmingham. In 1997 the government announced its decision. Can you predict what it was?

Who'd be a lorry driver on the M6? I could fly to Spain in the time it takes to get around Birmingham. If any of those road protesters tells me we don't need a new road I'll tell them what to do with their banner . . .

C

B Spaghetti Junction seen from a helicopter

Activities

1 Read the traffic report and look at map A.
  a) Draw a sketch map of the road network in the West Midlands. Start with the M6 like the example below. Add the other motorways to the map and label them. Then shade the urban areas on your map and label the main urban centres.

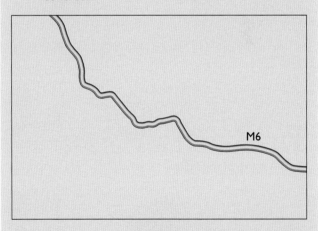

M6

  b) On your map, highlight the congested sections of road mentioned in the traffic report.

2 **a)** Use the map you drew in activity 1 to suggest the best route for a new road to relieve congestion on the M6. Mark your proposed route onto the map in pencil.
  b) Do you think that the government decided to build the road or not? Give reasons. You can find out what the decision was on page 40.

BUILDING BLOCKS

What are the arguments **for** . . .

In 1997 the government decided to go ahead with building the new road to relieve congestion on the M6. The route of the M6 toll, as it is known, is shown on map D. How does the route compare with the one that you drew on your map?

The M6 toll will be a 41 kilometre, three-lane motorway that cuts across the West Midlands linking junctions 4 and 12 of the M6. Long distance traffic using the M6 will be able to by-pass the congestion around Birmingham. The route goes through the **green belt** (land outside the urban area that is supposed to be protected from building development – see map H). A privately-owned company, Midland Expressway, has been asked by the government to build the road. In order to pay for the construction costs, and to limit the amount of traffic using the road, drivers will have to pay a toll. It will be the longest new road to be built in the UK for many years and is due to open in 2004.

> **Arguments in favour of new motorways**
> - Improve access and cut journey time for moving both people and goods
> - Traffic jams waste people's time, and cost industry in the UK £15 billion a year
> - Good roads are vital if a region wants to attract new industry and create jobs
> - Road building itself creates jobs
> - Benefit the environment by taking traffic off smaller roads and away from towns and cities
> - Fast-moving traffic on motorways produces less pollution than slow-moving traffic.

F Arguments for new motorways

Activities

1 Look at map D and map I.
   a) Compare the route of the M6 toll with the route that you drew on your map in activity 1 on page 39. Suggest why the government has chosen this route.
   b) From map I, describe the impact that the road will have on areas of protected environment.

2 Read the arguments for and against new motorways in tables F and G.
   Suggest how each of the following people would feel about the M6 toll. In each case choose at least one reason from the tables to support their opinion:
   a) a lorry driver on the M6 (see source C on page 39)
   b) a businessperson in the West Midlands (see source E)
   c) an **environmentalist** at Friends of the Earth (see source J)
   d) a village resident at Shenstone (see source H).

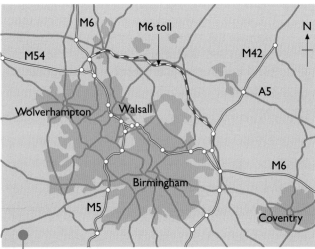

D The route of the M6 toll

E Tom Fanning, managing director of Midland Expressway

> Without the new road it will become harder for industry in the West Midlands to compete with other parts of the UK and Europe. Many firms will close or leave the region altogether. Transport problems add a huge amount to their costs.

... and *against* the new motorway?

Arguments against new motorways

- Encourage people to make car journeys that they would not have made otherwise
- Create more traffic that will eventually lead to the new motorway being overcrowded
- Damage the environment by taking farmland and woodland
- Attract industry which locates close to the motorway causing further environmental damage
- Encourage industry to locate away from the inner cities where jobs are needed
- Add to the problem of air pollution, noise and dirt.

G Arguments against new motorways

Assignment

You have to produce *either* a newspaper report *or* a TV news report about the new M6 toll. Aim to make it about 300 words, or three minutes long.

Imagine that the road is just about to open (if you are using this book after 2004 it should have opened already). You will have to interview people that are for and against the road to produce a balanced report. You can include quotations from the people in this Building Block or make some up.

For further help with your report look at the menu for writing about a controversial issue on page 63 in Unit 3.

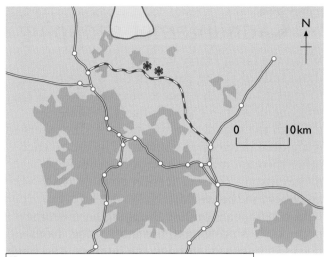

Key

- Urban area
- Green belt
- Area of Outstanding Natural Beauty (AONB)
- * Site of Special Scientific Interest (SSSI)

I Areas of protected environment on the M6 toll route

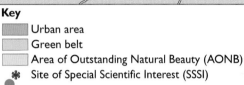

The M6 toll will cut through 41 kilometres of valuable green belt land and two Sites of Special Scientific Interest. This goes against the government's own policy, which is to put environmental protection before new roads. Even worse, it will fail to reduce congestion on the M6 because it will just create more traffic.

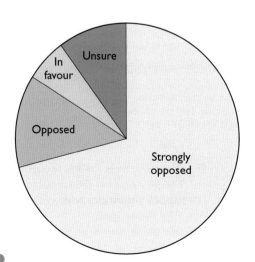

H Results of an opinion poll about the M6 toll carried out in the village of Shenstone

In favour

Unsure

Opposed

Strongly opposed

J Gerald Kells, environmentalist with Friends of the Earth

2.7 Can internet shopping help to make cars obsolete?

When the latest Harry Potter book went on sale thousands of young people in the UK were able to read it on the day that it was published. Were you one of them? This was made possible through the wonders of the **internet**. The internet is a global network of computers linked to each other through millions of telephone lines. It makes instant communication possible with anywhere in the world.

In recent years **on-line shopping** has become popular. It is possible to order anything, from books to bananas, through the internet, and have it delivered to your door (see source B). There has been a dramatic rise in the number of so-called **dot.com companies** that depend on the internet for their business. One of the best known of these companies is Amazon.com, a company that specialises in selling books.

A Another satisfied on-line customer!

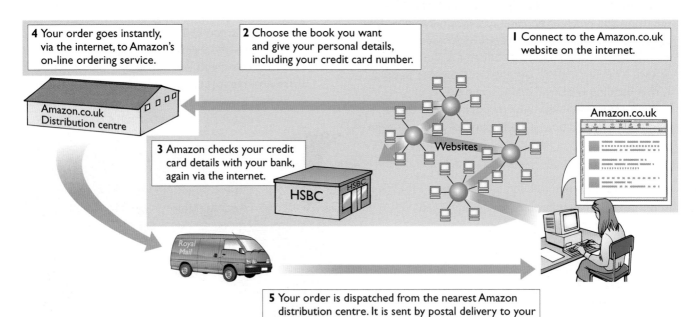

4 Your order goes instantly, via the internet, to Amazon's on-line ordering service.

2 Choose the book you want and give your personal details, including your credit card number.

1 Connect to the Amazon.co.uk website on the internet.

Amazon.co.uk Distribution centre

3 Amazon checks your credit card details with your bank, again via the internet.

HSBC

Websites

Amazon.co.uk

Royal Mail

5 Your order is dispatched from the nearest Amazon distribution centre. It is sent by postal delivery to your home and you could be reading it within 24 hours.

B How does on-line shopping work? The processes in the coloured panel all happen automatically via computers.

On the same day that thousands of happy young people were reading their Harry Potter book, many others were disappointed. They went shopping to buy the book, only to find that the shops had sold out. And it's not just books. Conventional shopping can be a frustrating and time-consuming experience, as you will know if you've ever looked for a parking space in a multi-storey car park on a Saturday afternoon! So, will the internet mean the end of shopping as we know it?

Amazon.com began in 1995 when the company's founder, Jeff Bezos, started taking orders for books over the internet on his computer at home in Seattle, USA. He wrapped the books in his garage and delivered them to the post office in the family car. Two years later Amazon supplied books to its one millionth customer! By 2001 the company had 29 million customers in 160 countries, making it the largest on-line shopping site in the world.

Population using the internet (%)

Key
1997
2001

D The rapid growth of the internet since 1997

C Inside Amazon's distribution centre in the UK

Share prices around the world fell again yesterday amid fears that the dot.com bubble has finally burst. The past six months have seen the collapse of several companies that, until just a year ago, were valued in millions of pounds. Even big players in the game are not immune from the sense of panic in the market. Amazon.com shares, worth $70 each a year ago, are now valued at less than $12. At the end of January the on-line retailer announced plans to cut 1,300 staff – 15% of its workforce – and warned that sales in the coming year would be 10% less than forecast.

E Has the dot.com bubble finally burst? 3 March 2001

Activities

1 Look at source B.
Describe in your own words how on-line shopping works.

2 a) What are the advantages for customers of on-line shopping over conventional shopping? Are there any disadvantages?
   b) Explain how on-line shopping would reduce the need for cars. Would it reduce traffic overall, do you think?
   c) Suggest what impact on-line shopping would have on:
      i)   other shops and shopping centres
      ii) employment
      iii) the environment.

Homework

3 Carry out a survey of people you know to find out their attitude to on-line shopping. Try to interview ten people, including people from different age groups.
   Do they use on-line shopping? If so, what do they buy? What are the benefits?
   If they don't use it, why not? Are they likely to start?
   Bring the results of your survey back to share with your class. What do the results of your survey suggest the future of on-line shopping will be?

4 ☐ Do you think there is a future in on-line shopping? Write a short essay of about 300 words to answer this question. Use the evidence on pages 42 to 43, together with the results of your survey to help you. The writing menu on page 63 in unit 3 may also help.

Could we live without cars?

OK, suppose we changed our shopping habits and started to do all our shopping on-line. Would it be possible to change the rest of our lifestyle and do without cars altogether? How would we get to work or to school? What about leisure and holidays? Would it be possible to do everything over the internet and, if it was, would we really want to?

At Crickhowell in Powys, Wales, a new village has been built for people who want a different lifestyle. It is called a **televillage**, especially built for people whose lives depend upon computers and who live and work under the same roof.

Every home in the village has a fast link to the internet. The whole idea is not so revolutionary as you might think. There are already two million **teleworkers** in the UK (almost ten per cent of the workforce) and the number is predicted to double over the next ten years.

Activity

Study the information about Crickhowell.
a) Would you want to live there? Give your reasons.
b) Why do you think that people here would you still need to use cars? (*Clue:* think about the size and location of the village.)

Crickhowell

F | The newly-built televillage at Crickhowell in Wales with the Brecon Beacons National Park in the background. Notice the cars.

Assignment

You are going to plan a town for the future where it will be possible to live without cars.

Before you start, think about these questions:

- How big should the town be?
- What shops and services would it need?
- How could it be planned to reduce the need to travel by car?
- Where would people live? Where would they work?
- How would people make their journeys?
- How could computers reduce the need for cars?

Design your town on a sheet of A4 paper. Choose a suitable scale so that it can fit on one sheet of paper. You could include any ideas that you have seen in this unit. Use your own ideas too. Be imaginative. After all, this is the future!

FOOTBALL AND FASHION –
Who are the winners in the global economy?

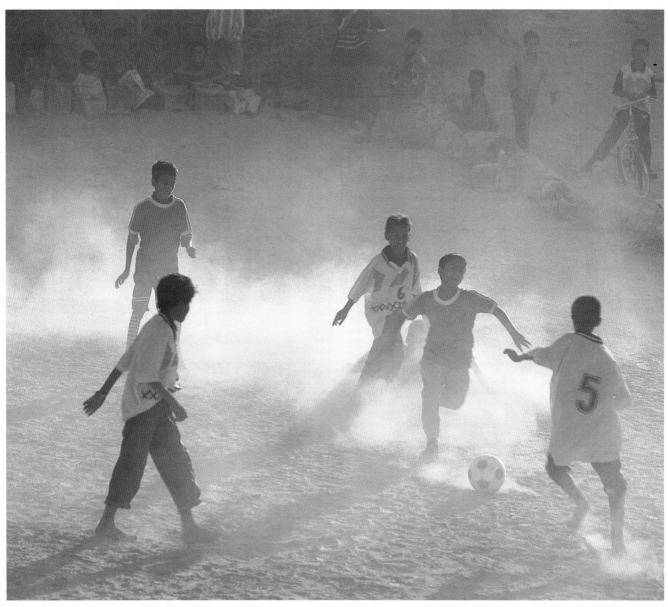

Football and fashion are both global **industries**. People all over the world play football, support the same teams and wear shirts made by the same companies. These boys are playing football in Yemen.

- If this photo had been taken in the UK what differences might you expect to see?
- How can you explain the differences?
- What questions would you like to ask the boys in the photo?
- Why are foreign football teams and their shirts so popular in countries like Yemen, do you think?

3.1 How are we connected with the world?

Geography is important! No matter where you live the **global economy** links you with every other part of the world. Goods and services produced in one place are sold and used somewhere else. Football and fashion are good examples of this.

The logos in source A belong to some of the world's biggest **brand** names in fashion. We buy their goods in our local shops, but they are produced on the other side of the globe. It's a similar story with football. People used to support their local team. These days they are just as likely to support a team in another city, or even in another country (map D).

A Do you recognise these logos ?

Brand name	Country the company is based in	Countries the products are made in
Adidas	Germany	Thailand ...
Ellesse	Italy	Korea ...
Fila	Italy	Vietnam ...
Le Coq Sportif	France	Indonesia ...
Nike	USA	Philippines ...
Reebok	USA	China ...
Umbro	UK	Bangladesh ...

B Some well-known brand names and their global connections

Activities

1 Look at sources A and B.
   a) Match the logos in source A with the brand names in table B.
   b) Make a large copy of table B, with an extra column to draw the logos. Leave space in the third column to add to the list of countries where their products are made.

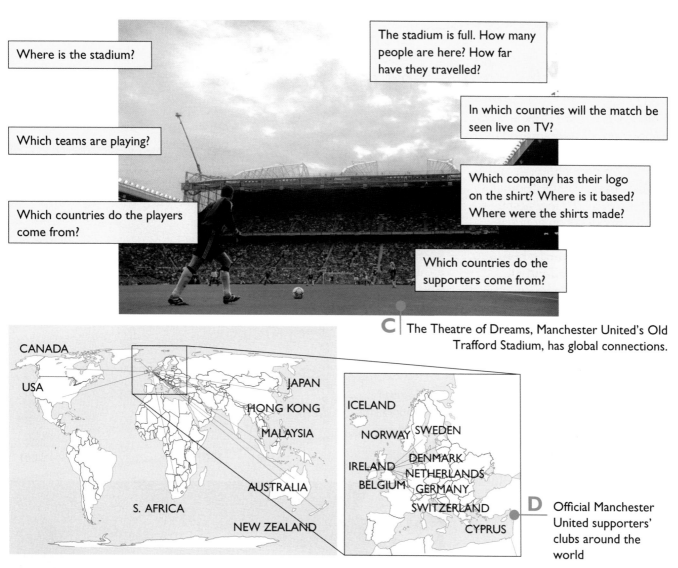

Where is the stadium?

The stadium is full. How many people are here? How far have they travelled?

Which teams are playing?

In which countries will the match be seen live on TV?

Which company has their logo on the shirt? Where is it based? Where were the shirts made?

Which countries do the players come from?

Which countries do the supporters come from?

C The Theatre of Dreams, Manchester United's Old Trafford Stadium, has global connections.

CANADA

USA

JAPAN

HONG KONG

MALAYSIA

AUSTRALIA

S. AFRICA

NEW ZEALAND

ICELAND

NORWAY SWEDEN

DENMARK

IRELAND NETHERLANDS

BELGIUM GERMANY

SWITZERLAND

CYPRUS

D Official Manchester United supporters' clubs around the world

Homework

2 Find out where the products for each brand name are made. You could investigate the labels on clothes and shoes in your own wardrobe. If you don't wear branded goods you could look in local shops. Make a list of the countries where they are made.
Share the information with your class. Add the information to the table you drew in activity 1.

3 Make a map of global fashion.
  a) On an outline map of the world, locate all the countries in table B (and the countries that you have added). You can use an atlas to help you.
  b) Use two colours to shade:
    i) countries where the companies are based
    ii) countries where the products are made.
  c) Describe the pattern on your map.

4 Look at photo C.
  a) Read the questions on the photo. Which questions can you answer from the information on these pages? Answer the questions.
  b) Which questions could you answer with a bit of extra research? Try the Manchester United website at www.manutd.com to answer the questions.
  c) Which questions are impossible to answer?

5 [ICT] Do some research on the internet to find out the global connections of your favourite team (if you haven't got one, do some research on your nearest team). Try to answer the questions in photo C for your team. Most football clubs now have their own website, so you could use that to help you.

FRAMEWORK

3.2 Football – a growing industry

Football is not the game it was. Most football stadiums today are unrecognisable from the way they were. Compare photos A and C.

Back in the 1980s, football in the UK reached a low point. Spectators stood on concrete terraces in half-empty stadiums. Violence was common both inside and outside football grounds and made people scared to go to games. Clubs were threatened with going out of business.

Everything changed after 1989, when 95 Liverpool fans were crushed to death at an FA Cup semi-final at the Hillsborough Stadium in Sheffield. The government decided that to avoid future disasters all major clubs must provide all-seater stadiums.

Football clubs were forced to modernise. The price of tickets had to go up to pay for the changes (graph B). Clubs that had been small family concerns turned into multi-million pound businesses. Football is now more than just a sport. It is an industry that employs thousands of people.

A Sunderland's old stadium before the club moved

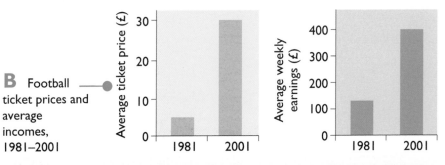

B Football ticket prices and average incomes, 1981–2001

C Sunderland's new Stadium of Light, 2001

Activities

1 Look at photos A and C.
   a) Describe the differences that you can see between the two photos.
   b) Explain why these changes happened.

2 Look at graph B.
   a) Describe how the change in ticket prices compared to the change in average incomes.
   b) Suggest how this affects the type of people that go to football matches.

3 Look at all the evidence on this page.
   a) How do you think these people feel about the changes in football over the past 20 years:
      i)   a football club director
      ii) a pensioner who has been a lifelong supporter
      iii) a family with three young children that like to go to football matches?
      In each case suggest why they may feel that way.
   b) Do you think the changes have been good or bad? Give reasons for your opinion.

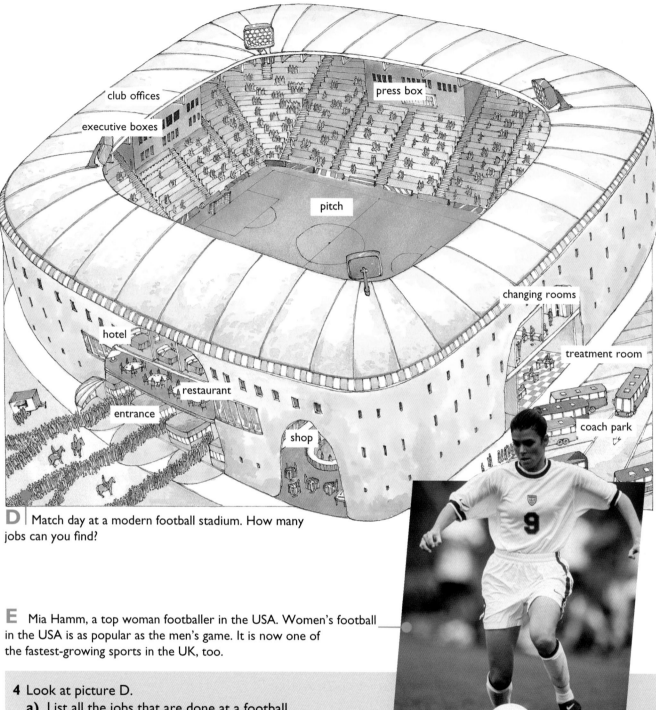

Labels on the stadium drawing:
club offices
press box
executive boxes
pitch
changing rooms
hotel
treatment room
restaurant
entrance
coach park
shop

D Match day at a modern football stadium. How many jobs can you find?

E Mia Hamm, a top woman footballer in the USA. Women's football in the USA is as popular as the men's game. It is now one of the fastest-growing sports in the UK, too.

4 Look at picture D.
  a) List all the jobs that are done at a football stadium. The places labelled on the drawing should give you clues.
  b) Think of jobs connected with football done *outside* the stadium (don't just think of match days). Make another list.
  c) Look at the list you made for activity 4a. Which of these jobs are more likely to be done by men? Which by women? (They could all be done by men or by women, but the question is who is *more likely* to do them.) How do you feel about this?

5 ⬜ Look at photo E.
Write a letter to your local football club to persuade them to do more to promote women's football. Explain how this could benefit:
  a) the football club
  b) girls and women.
You can use the writing menu on page 19 to help you to make your letter as persuasive as possible.

3.3 Do big cities always have the best teams?

Have you ever wondered why some football teams always do better than others? It's usually the big city teams with the largest stadiums that come top of the league. Look at the English Premier League table in source A and you will notice that all the teams are based in cities. Every successful team needs supporters to fill their stadium, so that they can afford to pay the players their huge wages. No supporters, no team!

But, is it quite so simple as that? Does it mean that the bigger the stadium and the bigger the city, the higher the team will come in the league? You are going to investigate the link between the position of teams in the Premier League and the size of their cities and stadiums.

Position	Team	Played	Won	Drawn	Lost	Points	Stadium capacity	City population
1	Manchester United	38	24	8	6	80	68,936	431,000
2	Arsenal*	38	20	10	8	70	38,500	6,962,000
3	Liverpool**	38	20	9	9	69	45,362	474,000
4	Leeds	38	20	8	10	68	40,204	724,000
5	Ipswich	38	20	6	12	66	22,600	114,000
6	Chelsea*	38	17	10	11	61	34,500	6,962,000
7	Sunderland	38	15	12	11	57	48,300	297,000
8	Aston Villa	38	13	15	10	54	43,250	1,008,000
9	Charlton*	38	14	10	14	52	20,043	6,962,000
10	Southampton	38	14	10	14	52	15,242	212,000
11	Newcastle	38	14	9	15	51	52,167	284,000
12	Tottenham*	38	13	10	15	49	36,257	6,962,000
13	Leicester	38	14	6	18	48	21,850	293,000
14	Middlesborough	38	9	15	14	42	35,100	147,000
15	West Ham*	38	10	12	16	42	26,054	6,962,000
16	Everton**	38	11	9	18	42	40,260	474,000
17	Derby	38	10	12	16	42	33,258	230,000
18	Manchester City	38	8	10	20	34	35,000	431,000
19	Coventry	38	8	10	20	34	23,627	302,000
20	Bradford	38	5	11	22	26	25,000	482,000

* Clubs in London   ** Clubs in Liverpool

A The end-of-season table for the English Premier League in 2000–2001. If you prefer, you could use the latest Premier League table to do the activities.

Activities

Work with a partner to complete Activities 1–3.
1 What do you think is the most important factor that determines a team's position in the league? Is it the capacity of their stadium (the number of people it holds)? Is it the population of the city? Or, are there other, more important, factors?

You are going to test the following two hypotheses:
- *The bigger its stadium, the higher the team will come in the league.*
- *The bigger the city it is in, the higher the team will come in the league.*

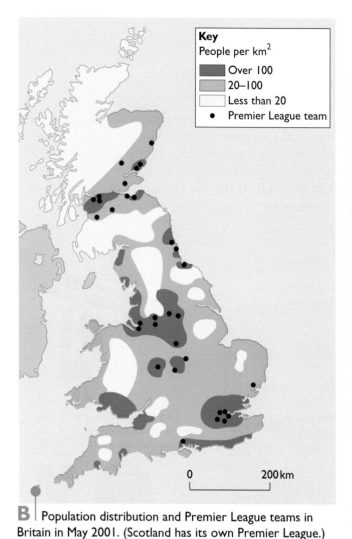

B Population distribution and Premier League teams in Britain in May 2001. (Scotland has its own Premier League.)

Key
People per km²
- Over 100
- 20–100
- Less than 20
- • Premier League team

3 a) You are going to draw two scattergraphs, one to compare the teams' league position with their stadium capacity and the other to compare it with the city population. You and your partner could draw one graph each.

To draw the graphs, draw a large grid like the one below. Mark each team according to its league position and its other rank. Two have been marked for you on the stadium capacity graph.

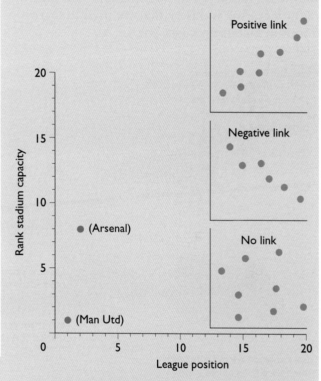

b) Look at the patterns on your graphs. Which of the three patterns shown are they most like? If the graph shows a positive link this would prove the hypothesis being tested, e.g. *The bigger its stadium, the higher a team will come in the league.* Write two paragraphs to explain whether the graphs help to prove, or disprove, the hypotheses.

2 You are going to test the hypotheses using table A. You could use the latest Premier League table if you prefer.
a) Rank the teams in order of their stadium capacity. Manchester United, with the largest stadium, will be ranked 1.
b) Rank the teams in order of the city's population. This is not quite so easy. Some cities have more than one team so, to be fair, you need to divide the population by the number of teams. For example, there are five Premier League teams in London, so you need to divide London's population by 5

$$\left(\frac{6,962,000}{5} = 1,392,000\right)$$

In this case the teams in London will be ranked equal first and the next team, Aston Villa (in Birmingham) will be ranked 6.

4 Look at map B.
a) Name the Premier League clubs on a copy of the map with the help of an atlas.
b) Compare the location of Premier League clubs with the population distribution on the map.
c) Name two areas that have high **population density** but no Premier League clubs. Name two Premier League clubs found in areas that do not have high population density.

FRAMEWORK

3.4 **B**rand new world of fashion

'What make is it?' That's the first question my kids ask when we go into a shop to buy clothes. Never mind the colour or whether it fits!

Like football, the fashion industry has changed. In the past, each company produced goods in their own **factory**. This is the traditional idea of manufacturing industry that we used to learn about in Geography (see source A).

But that is not the way things are now. **Transnational companies** (TNCs) operate all over the globe. They promote their brand image through never-ending advertising on TV, street billboards, and football shirts. We recognise these brands as easily as we recognise our own friends.

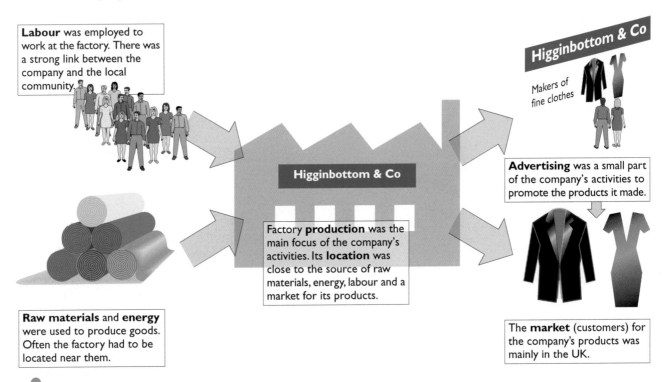

Labour was employed to work at the factory. There was a strong link between the company and the local community.

Higginbottom & Co

Factory **production** was the main focus of the company's activities. Its **location** was close to the source of raw materials, energy, labour and a market for its products.

Raw materials and **energy** were used to produce goods. Often the factory had to be located near them.

Higginbottom & Co

Makers of fine clothes

Advertising was a small part of the company's activities to promote the products it made.

The **market** (customers) for the company's products was mainly in the UK.

A The traditional clothing industry was based on factory production.

B An empty clothing factory in Newcastle upon Tyne

Activities

1 Look at source A and photo B.
   a) Draw a simple flow diagram to show how the traditional clothing industry worked. The diagram should include:

   raw materials, energy, labour, factory, advertising and market (customers).

   b) What benefits would this factory bring to:
   i) the local community   ii) the UK?
   c) What problems might have been caused when it closed?

Factories in the UK have been closing, so where are our clothes and shoes produced now?

Big brand-name companies find smaller companies in south-east Asia or central America to produce their goods. There, the cost of labour is much cheaper than it is in Europe or the USA. While the numbers of people employed in manufacturing in countries like the UK have fallen, they have grown in other parts of the world (table C). This is part of the process of **globalisation** (source D).

Country	Employment change (%)	Country	Employment change (%)
Finland	−71.7	Mauritius	+344.6
Sweden	−65.4	Indonesia	+177.4
Norway	−64.9	Morocco	+166.5
France	−45.4	Malaysia	+101.2
UK	−41.5	Mexico	+85.5
Germany	−40.2	China	+57.3
USA	−30.1	Turkey	+33.7

C Global changes in clothing manufacturing, 1980–93
Source: International Labour Office

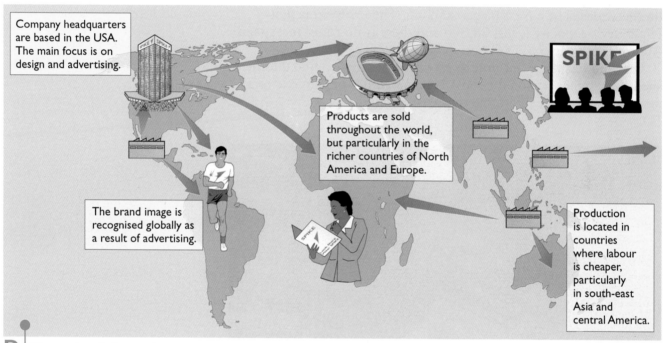

D The modern fashion industry is based on the brand image

2 Look at source D.
   a) Try to draw a flow diagram to show how the modern fashion industry works, similar to the one you drew for activity 1a.
   b) Explain why this was more difficult to draw.

3 Look at table C.
   a) Locate the countries on a world map. Shade the countries where manufacturing has increased in one colour and those where it has decreased in another colour.
   b) Describe the pattern of change in the fashion industry since 1980.

4 Compare the traditional clothing industry with the modern fashion industry. Which would be most likely to:

   • produce goods more cheaply
   • create more profits for companies
   • provide well-paid jobs
   • respond to the needs of customers
   • spend more on advertising?

In each case give a reason for your answer.

BUILDING BLOCKS

In this Building Block you will decide whether Arsenal should move to a new stadium and, if so, where it should be.

3.5

How did the Gunners decide where to go?

You may know already what the club decided but pretend you don't in case there is anyone who missed the news. The geography is the same whether the issue is past, present or future.

Arsenal is a football club with a long tradition. It was founded way back in 1886 and it has been playing at Highbury Stadium in the London Borough of Islington since the 1930s.

But, Arsenal (nicknamed the 'Gunners') believes that the stadium is no longer big enough to fulfil its ambitions. Over the past ten years it has been one of the most successful teams in the Premier League. It wants to compete with the biggest teams in Europe. To do this it needs to have a bigger stadium. The problem is that Highbury's capacity is just 38,000 and there is no room to expand. Almost every game is a sell out.

The club has three options: to stay at Highbury, to find another site nearby where it could build a bigger stadium or to move to a new location further from the centre of London.

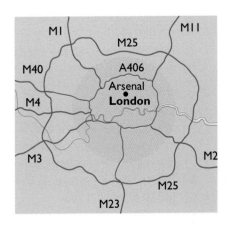

A An aerial view of Islington with Highbury Stadium at the centre

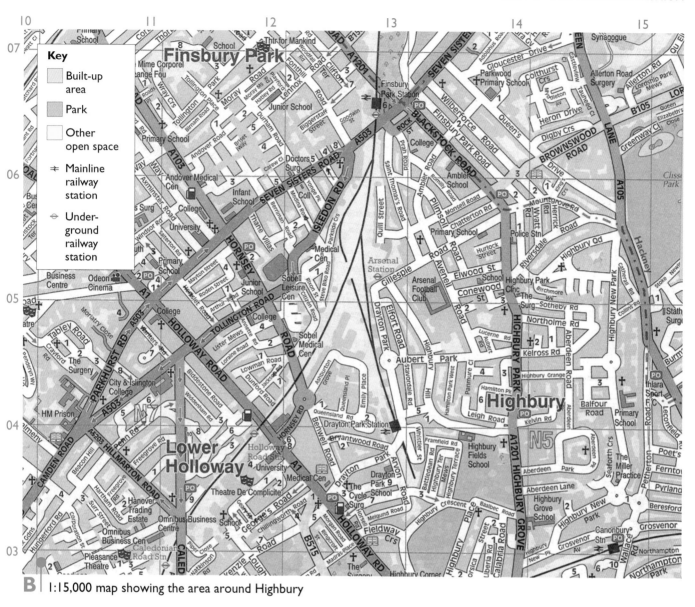

B | 1:15,000 map showing the area around Highbury

Activities

1 Look at photo A and map B. Find Arsenal's Highbury Stadium.

a) Suggest why the stadium was built here in the 1930s. (*Clue*: think about transport.)

b) Do you think that this is a good location for a football stadium today? Give reasons.

2 a) Draw an annotated map of the area in photo A. Your map should include the Highbury Stadium, roads, areas of housing, industry, and open space.

b) Annotate your map with the reasons you wrote for activity 1b to show whether or not it is a good location for a football stadium.

3 a) Locate four possible sites for a new stadium at the following grid references on map B: 107 045, 126 047, 136 038, 148 055.

b) Consider whether any of these sites might be suitable. You should think about:

- space available for the stadium (needs at least double the size of existing stadium)
- how the space is used now
- buildings that would have to be demolished
- access by public transport and by car
- the impact of the stadium and traffic on neighbouring residents.

c) Identify the best local site for a new stadium.

Inner city or out-of-town?

C Match day outside Arsenal's stadium in Avenell Road (where Nick Hornby lived when he wrote *Fever Pitch*)

Most football clubs in the UK were started about 100 years ago when cities were growing rapidly. As a result many football stadiums are located in the midst of densely-populated inner city areas. Highbury, hidden away among rows of tightly-packed terraced houses, is a good example. Because road access to the stadium is so difficult, 70 per cent of Arsenal fans come to games by public transport – the highest percentage in the Premier League.

During the 1990s, some clubs moved to out-of-town sites where there is space to build bigger stadiums with more car parking. These also have better road access via the motorway network. Arsenal has considered moving to an out-of-town site, further from the centre of London.

Nobody supports Arsenal in my street. Some of my neighbours are what used to be known, years ago, as yuppies, and they have no interest in football; others are transients, squatters or short-lease tenants, never around for long enough to acquire the taste for it. The rest of them . . . I don't know. You can't come up with a theory for everyone, and there's no accounting for taste.

I suspect that I moved here a good twenty years too late, and that for the last couple of decades the local support has dwindled away steadily. According to the club's information, a huge percentage of fans live in the Home Counties (when I travelled down from Cambridge, the trains were packed with Arsenal supporters by the time we got to Hatfield). The demographics have changed now, and all those people who used to walk to the game from Islington and Finsbury Park and Stoke Newington have gone: they're either dead or they've sold up and moved out to Essex or Hertfordshire or Middlesex.

I'm more alone here than I ever thought I would be at the end of the sixties, all those years ago, when I used to pester my dad to buy a house on Avenell Road, and he said I'd get fed up with it.

D Extract from *Fever Pitch*, Nick Hornby's novel about a lifelong Arsenal supporter

Activities

1 Look at photo C.
  Imagine that you are in the middle of the crowd.
  – What can you see?
  – What can you hear?
  – What can you smell?
  – How does it make you feel?

  L Write a short story or a poem about going to your first football match. If you've never been to a football match, even better – you can use your imagination!

2 Read extract D.
  a) Why do you think that Nick Hornby always wanted to live on Avenell Road?
  b) Do you think that he will eventually get fed up with it? Give reasons.

New football stadiums shouldn't be built near people's homes. We are worried about Arsenal's proposal to build a large new stadium at the end of our road. An inner city residential area is a ridiculous place to build a stadium of this size. There are going to be even more problems with car congestion, noise and litter than we have already. What we want to know is how the club expects 60,000 people to get to matches without causing more disruption for us.

E Nick Robinson, leader of the local resident's action group

It is really important for the club and for the fans. At Highbury you can fit 38,000 people in, and I'm sure more want to come every week. With the new stadium they could. Arsenal could become one of the biggest clubs in Europe. If they want to have a chance of that then they have to have a bigger stadium.

F Patrick Vieira, one of Arsenal's players

Decision-making activity

Work with a partner. You will represent the board of directors at Arsenal Football Club. You have to make a decision about where the club should go. You have three main options:

A to stay at Highbury

B to build a bigger stadium at a local site in Islington (this could be the one that you identified in activity 3 on page 55)

C to move to a new stadium further from the centre of London.

You will need to consider the following factors before you make your decision:

- the cost of land (generally cheaper as you move away from the centre of London)
- building costs (the more buildings that have to be demolished, the higher these will be)
- space for expansion
- access by road
- access by public transport
- environmental impact (particularly noise and disturbance for residents)
- local support and tradition.

Are there any other factors that you need to consider?

Decision-making factors	Stay at Highbury	Build new local stadium	Move further from London
_____ ×7			
_____ ×6			
_____ ×5			
_____ ×4			
_____ ×3			
_____ ×2			
_____ ×1			
Total score			

N Work with your partner.

a) Rank the factors in order of their importance. On a copy of the table above, write the factors in the first column in order of importance.

b) Give each option a score of 1, 2 or 3, depending on whether it is worst, better or best for each factor. Write the scores in the top left corner of each box.

c) Multiply each score by the number on that row to obtain a weighted score. Write these scores in the bottom right corner of each box.

d) Add the weighted scores to obtain a total score for each option.

Which option would you choose?

Will the Reds go Green?

Arsenal has decided to build a new stadium at a local site on Ashburton Grove, less than a kilometre from Highbury (map G). Did you decide on a new local stadium in the decision-making activity? The site is the one at grid reference 126 047 on map B on page 55. Is this the same one that you identified?

However, one important obstacle still remains. Arsenal must obtain planning permission from Islington Council before it can build the stadium. One factor that the council must consider is the impact that the stadium will have on the environment.

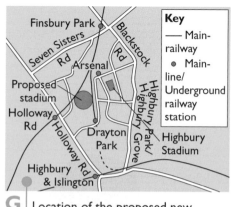

G | Location of the proposed new stadium at Ashburton Grove

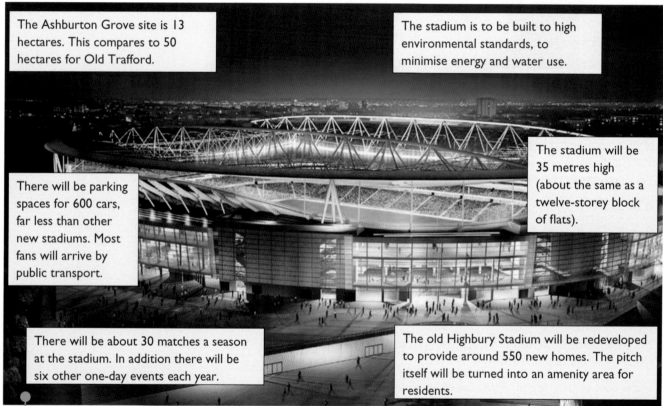

The Ashburton Grove site is 13 hectares. This compares to 50 hectares for Old Trafford.

The stadium is to be built to high environmental standards, to minimise energy and water use.

There will be parking spaces for 600 cars, far less than other new stadiums. Most fans will arrive by public transport.

The stadium will be 35 metres high (about the same as a twelve-storey block of flats).

There will be about 30 matches a season at the stadium. In addition there will be six other one-day events each year.

The old Highbury Stadium will be redeveloped to provide around 550 new homes. The pitch itself will be turned into an amenity area for residents.

H | A computer-generated image of the new stadium

Assignment

Work in a small group.

1 You represent the planning committee for Islington Council. You have received the planning application from Arsenal Football Club to build its new stadium at Ashburton Grove. Study the above proposals carefully.

2 Individually, decide whether you would give planning permission or not. Discuss the planning application with your group. Explain your own point of view and listen to other members of the group explain theirs. Then come to a decision as a group.

If you decide to give permission, what conditions would you require the club to meet? Think particularly about the impact of the stadium on the environment.

If you decide to refuse permission, how could Arsenal improve the application if it wants to apply again?

3 [L] Write a letter to Arsenal Football Club to explain the planning committee's decision.

In this Building Block you will consider the arguments for and against buying branded clothes and write an article for a fashion magazine.

3.6
Should we buy the logo?

I can't believe they need to charge so much for these trainers. That's about a month's wages on my paper round!

It can't be because they're expensive to make. They pay their workers peanuts, I've heard.

The best thing is not to buy anything with a logo on. Then the companies would have to listen.

You'll never beat the big brands. They know that we're not cool if we don't wear their logos.

But if we didn't buy the clothes, then those poor people would get no money at all.

A Even shopping can be a geography lesson!

Activity
Look at photo A.
  a) Read the statements. Discuss them with a partner. Which do you agree with and which do you disagree with?
  b) Make copies of the statements. Put them in order, from the one that you agree with most, to the one that you disagree with most. Keep this until the end of the Building Block, then see if your ideas have changed.

Should we buy the logo? The case against . . .

A century ago in the UK, clothes were made in small factories and workshops dotted around our big cities. Sometimes these were known as **sweatshops** because of the poor conditions that people worked in. Gradually, the sweatshops disappeared as workers fought for better wages and working conditions. But now sweatshops have reappeared, this time dotted around the globe in the places where our clothes are made today.

Over the past twenty years, trans-national companies have moved production of clothes and footwear to countries where labour is cheap and laws about working conditions are less strict. In particular, they have favoured the **newly-industrialising countries** of Central America and South-east Asia. China is now the world's largest clothes producer: in 1996, 44,000 clothes factories in China employed four million workers, most of them women.

Country	Hourly wage rate in 1998 (US$)
USA	$8.42
UK	$7.60
Philippines	$0.62
El Salvador	$0.60
Mexico	$0.54
Honduras	$0.43
China	$0.30
Nicaragua	$0.25
Indonesia	$0.22
India	$0.20
Bangladesh	$0.17

B Hourly wage rates in the clothing industry in selected countries

MAQUILA Solidarity Network maquilasolidarity.org

Meet Yin

Yin works in a garment factory in Shenzhen, a 'special economic zone' in Guangdong Province southern China (see map A on page 114). She sews clothes for well-known North American labels.

Workers such as Yin are known as 'mingong', or peasant labourers. She left her rural village in search of work at the age of seventeen. She had to get a temporary resident's permit to be able to work in Guangdong. The company helped her to get the permit and then took deductions from her pay cheques over the next six months. They also keep her resident's permit, making it impossible for her to leave the factory grounds without permission.

Yin and other workers in the factory live in a dormitory behind the plant. She sleeps in a room with eleven other women sharing double beds. Each woman has about one square metre of living space. For as long as she works here, this will be her home. The cost for her meals in the dorm is deducted from her wages.

Yin doesn't spend much time in the dorm. She has only two days off a month and regularly works fourteen hours a day. Her work starts at 7.30 in the morning and usually ends at 9.30 at night. During peak production periods she sometimes works through the night. According to Chinese labour law, Yin is entitled to time-and-a-half for overtime after eight hours and double time on Saturday and Sunday. But Yin and her co-workers aren't aware of what the law says. She never gets time-and-a-half for her overtime and regularly works more than the legal limit of overtime hours. The money she earns helps to support her family back home in the village.

Discipline in the factory is strict and workers are fined for any violation of factory rules. There are fines for talking or getting a drink of water during work time, for arriving late, for refusing to work overtime, and for cooking in the dorms. If Yin misses three days of work in a row the company will consider her to have resigned and she will have no job to return to.

Despite the sweatshop conditions at work, Yin's biggest worry is that the factory will close down and she'll be left without a job. Many companies are leaving Shenzhen to move farther north where local governments are offering investment incentives and labour is cheaper.

C A worker's story from the Maquila Solidarity Network website. (*Maquila* is a word used for a workshop or factory in Central America.)

... and the case in favour

Companies like Nike and Adidas have been stung by criticism that they exploit workers in poorer countries. It is certainly not cool to be seen to be employing child labour, for example. Anything that damages their brand image may reduce sales – no matter how good their clothes are. Nike, Adidas and other companies now have rules to control working conditions in the factories that produce their goods.

There has been a backlash against these companies, particularly in America. College sports teams – sponsored by companies in the way that football teams are in the UK – have changed their sponsors. This has hit the companies' reputations and their sales. Nike has responded by putting out information on its website to restore its image.

D Phil Knight founded Nike in the 1970s with an idea for a sports shoe. By 1998, the company was worth $8 billion.

Some frequently asked questions about Nike

Does Nike use child labour?

How much do workers earn in factories?

Nike makes a lot of profits. Phil Knight is a multi-billionnaire. Nike athletes like Michael Jordan and Tiger Woods earn millions from Nike. Given that, how does Nike justify what it pays its workers?

Don't you just move your production from country to country, searching for the cheapest labour?

E These questions appear on the Nike website: Nikebiz.com

Activities

1 Look at table B. Which of the countries in the table were your clothes made in? (see activity 2 on page 47)? How do you feel about this?

2 a) Visit the Nike website at Nikebiz.com. Find the answers to the frequently asked questions in the 'responsibility' section under 'labor'. Print out a copy so that you can do activity 3.

b) Why do you think Nike have these types of questions and answers on their website?

3 ⬛ Work with a partner. One of you should read extract C and the other should read the answers to the questions in box E from the Nike website.

Make notes to summarise the main points in each source. From C, note all the ways in which workers are treated unfairly. From E, note all the ways in which workers around the world benefit from Nike.

Share the notes you made with your partner. The easiest way may be to make a photocopy.

4 Still working with your partner, role play an imaginary conversation between Yin and Phil Knight.

If you read extract C, imagine that you are Yin. Think of at least five questions that you would like to ask Phil Knight.

If you read box E, imagine that you are Phil Knight. How would you answer Yin's questions?

Just think about it!

Geography textbooks can't tell you what to do (teachers and parents do that!). But they can encourage you to think. Trans-national companies, like Nike, spend huge amounts of money on advertising. That obviously has an important influence on the way we think. If we want to find a different point of view we have to look further. Organisations like Maquila Solidarity Network try to persuade us not to buy products from companies that continue to exploit their workers.

Which of them is right? If you're not sure, source F presents a range of facts from both sides of the argument to help you to think about it.

Activity

What do you think now?
Look back at the statements that you read on page 59. Do you agree or disagree with them now? Would you still put the statements in the same order? Have your ideas changed? In what way?

F Arguments for and against buying products with a logo

TNCs provide employment for people in poorer countries who otherwise would have no work.	Workers for TNCs in poorer countries earn about ten per cent of someone doing a similar job in the UK.	The workers who produce clothes and footwear for TNCs cannot afford to buy the products themselves.
TNCs have no commitment to a particular country. They can relocate their business wherever they want.	Investment by TNCs in poorer countries leads to factory closure and creates unemployment in richer countries.	TNCs now allow independent monitors into factories to ensure that they don't employ children.
TNCs advertise to persuade people to buy products that many of them may not really be able to afford.	Wages for TNC workers can be 25 per cent higher than the minimum wage in poorer countries.	Workers in many factories have to endure poor working conditions and abuse of their human rights.
TNCs sell products that many people want to wear.	Wages in poorer countries may not be as low as they seem, because the cost of living is often much lower than it is here.	Less than five per cent of the price you pay goes to the person that made the product. Fifty per cent goes to the shop where you bought it.
Wages in poorer countries enable workers to support their families, who often live in the poorest rural areas.	The governments of many poorer countries do not permit the formation of trades unions to protect workers' rights.	TNCs sponsor many community activities in richer countries. Nike gives money to organisations working in poor, black communities in the USA.

Assignment

You are going to write an article for a fashion magazine, to advise young people about buying branded clothes and footwear. You have to write a balanced article that presents both sides of the argument.

Aim to write about 500 words, but it could be more or less than this. Remember that you are writing for a young audience, so write in a style that will appeal to them. The menu on the opposite page will give you ideas to help you with your writing.

Before you write, you need to think about the arguments that you will include. Use source F to help you. Cut out copies of the arguments. Sort them into three groups – **economic, social** and **ethical**. You could divide your article into three sections dealing with the three types of argument.

Now divide each group into arguments for, and arguments against, buying products with a logo. Each section within your article could be divided into two paragraphs using these arguments.

How to write about a controversial issue

Text type	Tense	Starters	Links	Conclusions	Vocabulary
Recount	Past	First, …	… but …	In conclusion, …	industry
Description	*Present*	Second, …	… yet …	In summary, …	global economy
Method	Future	In the first place, …	… whereas …	Overall, …	globalisation
Explanation		More importantly, …	… while …	On the whole, …	trans-national company
Persuasion		For example, …	However, …	In short, …	factory
Discussion		For instance, …	On the other hand, …	In brief, …	employment
		Finally, …	In contrast, …		product
			Although …		advertising
			In spite of …		labour
			Alternatively …		sweatshop
					exploit
					trade union

STRUCTURE FOR WRITING ABOUT A CONTROVERSY

1 The title could be a question that summarises the controversy. For example, *Should we buy the logo?*

2 The first paragraph should introduce the controversy. You need to preview the main points on both sides of the argument and say who represents each viewpoint. Write in the present tense and use the third person. For example, *Companies like Nike argue that . . .*

3 The main part of your writing should present all the arguments on each side of the controversy, together with all the supporting evidence. It helps if you can break this into sections dealing with different types of argument. For example, *First, let's look at the economic arguments in favour of . . .*

Each section should present the arguments for and against. It may help if you keep all arguments on each side of the controversy together – a paragraph on the arguments in favour and then a paragraph on the arguments against. Alternatively, you can go through all the arguments and counter-arguments one by one. For example, *Trans-national companies say that they bring economic benefits to poorer countries . . . whereas organisations like Maquila Solidarity Network believe that TNCs have no commitment to poorer countries.*

4 Write a final paragraph to come to your conclusions. Weigh up the strength of the arguments on each side. If you want you can change from the third person to the first person to give your own views. For example, *The next time I go into a shop to buy a pair of trainers, I would . . .* Don't worry if you can't make up your own mind. The important thing is that you make your readers think about the arguments.

GROUNDWORK

FRAMEWORK

BUILDING BLOCKS

DIGGING DEEPER

3.1 Who are the real Olympic champions?

At Sydney 2000, Great Britain had its best Olympic Games since 1912. It won eleven gold medals, ten silver and seven bronze, ranking it tenth in the medal table (table B). The USA came top of the table, but then it would, wouldn't it!

You would expect large wealthy countries to do well in the Olympics. They have more people to choose from and more money to help them to train. Three of the four most populated countries – China, Russia and the USA – were in the top three medal places. Only India was missing. But look which country comes in at number twenty – Ethiopia! It is one of the poorest countries in the world, and it has a similar population to ours. Obviously, the Olympic medal table is not as predictable as we thought.

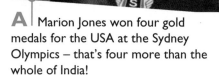

A Marion Jones won four gold medals for the USA at the Sydney Olympics – that's four more than the whole of India!

Rank	Country	Gold	Silver	Bronze	Score	GNP ($ billion)	Alternative score	Alternative rank
1	USA	39	25	33	200	7,904	0.025	
2	Russia	32	28	28		907		
3	China	28	16	15		3,779		
4	Australia	16	25	17		409		
5	Germany	14	17	26		1,807		
6	France	13	14	11		1,248		
7	Italy	13	8	13		1,173		
8	Netherlands	12	9	4		350		
9	Cuba	11	11	7		19		
10	Great Britain	11	10	7		1,200		
11	Romania	11	6	9		125		
12	South Korea	8	9	11		616		
13	Hungary	8	6	3		99		
14	Poland	6	5	4		292		
15	Japan	5	8	5		2,982		
16	Bulgaria	5	6	2		39		
17	Greece	4	6	3		147		
18	Sweden	4	5	3		176		
19	Norway	4	3	3		116		
20	Ethiopia	4	1	3		35		

B The top twenty countries at Sydney 2000 – based on the number of medals they won

Richer country

Poorer country

GNP, or **gross national product**, is the total value of all the goods and services produced each year in a country. It reflects both the population and the wealth in a country. The more people there are, or the more each person produces, the higher GNP will be. But, a word of warning: *GNP does not measure wealth*. It is possible for a country to have a high GNP, but for its people to be poor. Can you think how? (Cartoon C gives you a clue.)

Wealth in a country is measured using **GNP per capita**, that is gross national product divided by the number of people. GNP per capita varies around the world, as you can see in map D.

C GNP is not the same as wealth!

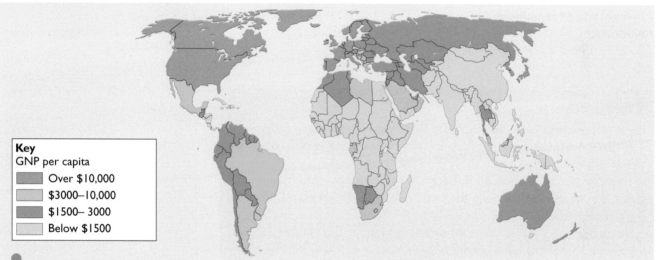

Key
GNP per capita

- Over $10,000
- $3000–10,000
- $1500– 3000
- Below $1500

D GNP per capita by country

Activities

1 You are going to use a copy of table B to devise an alternative Olympic league table. It will measure each country's relative sporting achievement, using GNP to take its population and wealth into account.

   a) Calculate a medal score for each country. Give three points for each gold medal, two for each silver and one for each bronze. Add the points together and write the score in the sixth column. The USA is done for you.

   b) Calculate the alternative score for each country. Do this by dividing the score by GNP. This will give you the number of points per $ billion of GNP. You can use a calculator to do this. Write the alternative score in the eighth column. Again, the USA is done for you.

   c) Finally, in the last column, rank the countries according to their alternative scores.

2 Compare the two rank orders: the official Olympic rank order according to the number of medals in table B, and your alternative rank order.

   a) Which provides the best measure of countries' sporting achievement, do you think? Why?

   b) What surprises you about your alternative rank order? Why?

3 Using an atlas, locate the countries in table B on map D.

   a) Where are most of the countries in the official top twenty located? Explain why.

   b) Where do you think most of the countries outside the top twenty would be located? (You can check this from an extra sheet that your teacher may give you.)

   c) How do you think your alternative rank order would change the pattern?

Turning the medal tables?

CUBA

Population:	11,059,000
Size:	111,000 km²
GNP per capita:	$1,170
Life expectancy:	76
Adult literacy:	96%
Infant mortality:	7 per 1,000
Access to safe water:	93%
Primary school enrolment:	100%
Percentage of females in workforce:	43%

Cuba is the largest island in the Caribbean. For hundreds of years it was a Spanish colony. Later it came to be dominated by the USA. In 1959 it had a revolution that brought a socialist government to power. It made health and education its priority for the country's development. There are few wealthy people in Cuba but there is not much poverty either. However, this may change as Cuba has begun to open its economy to the outside world. It has recently become an important tourist destination.

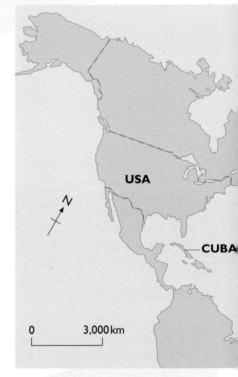

Assignment

Plan an investigation to find out why Cuba did so well in the Olympic Games, and why, taking into account its population and wealth, the USA did so poorly.

Look at the information on this page. It may give you some clues about the differences between the two countries and get you thinking in the right directions.

Think about questions you need to ask to find out the reasons for the difference in sporting achievement between the two countries. For example: *Are Cubans naturally better at sport? Do more people have the opportunity to play sports in Cuba? How could health and education affect sporting performance?* There are many other questions you might think of.

Then think about how you could carry out the investigation. What information would you need to obtain? How would the internet be able to help you?

You may not have time to carry out your investigation. The aim of this assignment is to make you *think* about the questions you need to ask and *how* you would investigate them. But if you have a long holiday coming up . . . !

The **USA** is one of the wealthiest countries in the world and it has the world's largest economy. It has a democratic government that promotes development based on a free market economy. In the USA some individuals and companies have become very wealthy, but poverty is also widespread. Public services like health and education have been cut back. People have to pay for their healthcare through private insurance and not everyone can afford to go to college.

USA

Population:	267,636,000
Size:	9,159,000 km²
GNP per capita:	$29,080
Life expectancy:	76
Adult literacy:	99%
Infant mortality:	7 per 1,000
Access to safe water:	90%
Primary school enrolment:	96%
Percentage of females in workforce:	46%

There are few places left on our planet that remain unchanged by people. Even the most remote corners of the world are now threatened by human activities.

- Where do you think this photo was taken?
- What are people doing here?
- In what ways could they be a threat to the environment?
- What might the penguins be thinking?

GROUNDWORK

4.1 What's your picture of the countryside?

Do you know this painting? You can find it hanging in living rooms all over the country (not the original – that's in the National Gallery!) Constable is still one of our most popular artists, even though he was painting almost 200 years ago. His paintings portray an image of the English countryside that we all recognise.

A century later, the writer Laurie Lee described a very similar landscape (extract B). Even today you can still find images like this of the countryside, anywhere from a holiday brochure to the plastic wrapper on a loaf of bread!

Of course, in real life this countryside no longer exists even if we like to think it does. Source C shows other images that people living in cities have of the countryside today.

A *The Cornfield* painted by the English artist, Constable, in the early nineteenth century. It shows the Suffolk landscape of his childhood.

The stooping figure of my mother, waist-deep in the grass and caught there like a piece of sheep's wool, was the last I saw of my country home as I left it to discover the world. She stood old and bent at the top of the bank, silently watching me go, one gnarled red hand raised in farewell and blessing, not questioning why I went. At the bend of the road I looked back again and saw the gold light die behind her; then I turned the corner, passed the village school, and closed that part of my life for ever.

It was 1934. I was nineteen years old, still soft at the edges, but with a confident belief in good fortune. I carried a small rolled-up tent, a violin in a blanket, a change of clothes, a tin of treacle biscuits, and some cheese. I was excited, vain-glorious, knowing that I had far to go; but not, as yet, how far. As I left home that morning and walked away from the sleeping village, it never occurred to me that others had done this before me.

I was propelled, of course, by the traditional forces that had sent so many along this road – by the small tight valley closing in around one, stifling the breath with its mossy mouth, the cottage walls narrowing like the arms of an iron maiden, the local girls whispering, 'Marry, and settle down.' Months of restless unease, leading to this inevitable moment, had been spent wandering about the hills, mournfully whistling, and watching the high open fields stepping away eastwards under gigantic clouds . . .

B The opening paragraphs from Laurie Lee's autobiography, *As I Walked Out One Midsummer Morning*

C | Modern images of the countryside

Activities

1 Look at picture A. Analyse the painting in the same way you would a photograph.

- What does it tell you about the place?
- What can you guess about the place?
- What doesn't it tell you about the place?
- What do you think this place might look like today?

What other questions would you like to ask?

2 Read extract B.

a) Identify ten words or phrases that describe the traditional countryside in which Laurie Lee was brought up.

b) What do you think were the factors that made him leave home (push factors)?

c) Would you have made the same decision? What are the push factors that would make you go, or the pull factors that would make you stay?

3 Work with a partner.

a) If you live in a city, brainstorm your ideas of what the countryside is like. Compare them with the ideas in C. What other ideas did you have?

If you live in the countryside, how accurate do you think these images in C are? Which of them would you choose to describe the area where you live?

b) Identify which ideas are positive and which are negative. On balance, do you have a positive or negative image of the countryside?

c) Would you rather live in the countryside or in the city? Give your reasons.

4.2 Our disappearing countryside

Look at drawing A. It shows things you would find in the British countryside today. For 200 years people like the young Laurie Lee left the countryside to live in cities. This process was called **urbanisation**. But now the process has begun to reverse (graph B). People are moving back to live in the countryside. And if they can't live there, they want to be able to visit.

Rural areas now provide land for roads, housing and recreation to ease the pressure on land in the cities. This process of **counter-urbanisation** is changing the whole character of the countryside. Every year an area of countryside the size of Bristol is built on. At this rate, by 2050 twenty per cent of the countryside in England will be covered in concrete.

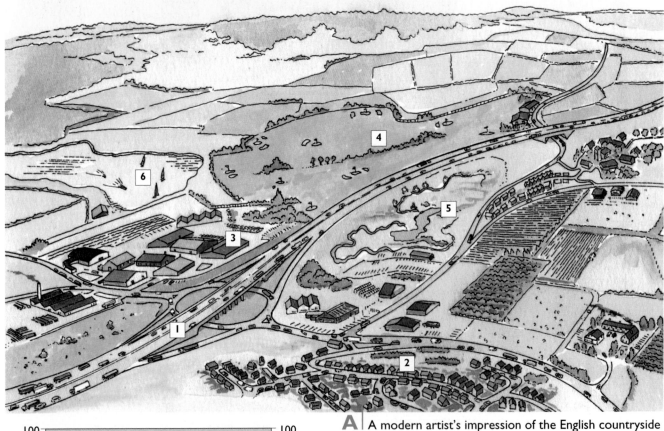

A A modern artist's impression of the English countryside

B Changes in the balance of urban/rural population in the UK

People in cities think of the countryside as somewhere to get way from it all – a place to find tranquillity. This will no longer be possible if the rural environment continues to change. The maps in C show that **tranquil areas** have shrunk as a proportion of the total area of England since the 1960s.

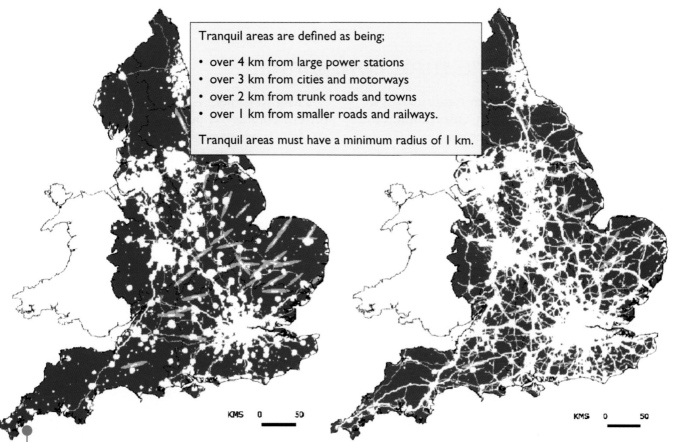

Tranquil areas are defined as being;

- over 4 km from large power stations
- over 3 km from cities and motorways
- over 2 km from trunk roads and towns
- over 1 km from smaller roads and railways.

Tranquil areas must have a minimum radius of 1 km.

C Tranquil areas (shaded in green) in England in the 1960s (left) and the 1990s (right), according to the Council for the Protection of Rural England (CPRE)

Activities

1 Look at drawing A. How has the countryside changed since the nineteenth century? Look back at source A on page 68 for some ideas.
   a) Identify six things that you can find in the countryside today that you would not find in a Constable painting. The numbers will give you clues.
   b) Describe the changes that might have happened to the following features in the drawing:
      i)   farming
      ii) the village
      iii) the barn.

2 Look at graph B.
   a) Describe the changes in the urban and rural population in the UK since 1801. When did rural population stop falling?
   b) Explain why the changes happened.

3 Look at the maps in C.
   a) Describe carefully the changes that have happened in the countryside since the 1960s. In which parts of England have most changes happened?
   b) Compare the maps with map A on page 26. What do you notice?
   c) Suggest the main reasons that the tranquil areas of England have shrunk.

Homework

4 Look at a map of your local area (a 1:50,000 Ordnance Survey map would be ideal). If you live in a city, the map should include the surrounding countryside. Trace the main urban areas and transport routes to make your own map. Using the definition of a tranquil area in source C, work out where the tranquil areas would be. (Remember 2 cm on a 1:50,000 map is the equivalent of 1 km in real life.) Shade the tranquil areas on your map.

4.3 Protecting the rural environment

As pressure on land in the countryside grows, so does the need to look after it. **Conservation** is the way that we protect the environment.

In the UK, all sorts of people are involved in conservation, from the government, who decides which areas to protect and makes the laws, down to local voluntary groups, who do jobs like clearing footpaths and restoring ponds. In between the government and local voluntary groups are a range of other organisations like the National Trust, the Royal Society for the Protection of Birds (RSPB) and Friends of the Earth. Most of them depend on donations and volunteers to keep them going. Do you feel strongly about the environment? If you don't already belong to one of these organisations then perhaps you could get involved.

The government has established areas where the environment is protected. Can you recognise the four types of protected area in photos A–D?

- **National parks** are large areas of beautiful countryside where people go for recreation, but where other people still live and work.
- **Heritage Coast** is stretches of beautiful coastline that are protected in a similar way to national parks.
- **Environmentally Sensitive Areas** are areas where the beauty of the landscape depends on maintaining traditional farming methods.
- **Nature reserves** are small areas where plants and animals need to be protected from human activities that could disturb, or destroy, them.

A

C

B

D

The main rural activity is still farming. Traditional farming methods created **habitats** where many types of wildlife thrived. Birds like the skylark used to be common in the UK. The skylark nests in crops and short grass during the summer. Modern **intensive farming** methods destroy suitable nesting sites. Source E shows that early harvesting of winter-sown crops and regular cutting of grass for silage disturbs the skylark during its nesting period. As a result its population has fallen by 60 per cent over the past 25 years. The loss of birds from the countryside has become a big worry for conservationists.

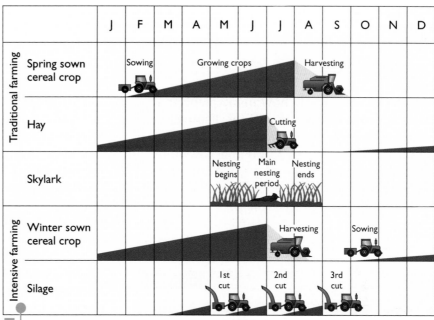

E How traditional and intensive farming methods affect the skylark

F A skylark

Activities

1 You are going to complete a large table to compare four types of protected area.
   a) Match photos A–D to one type of protected area. In the first column of your table, list the letters A–D and, in the next column, write the types of protected areas.

Photo	Type of protection	Description	Why protect it
A			

   b) In the third column, write a sentence to describe the environment in each photo.
   c) In the final column, suggest why each environment should be protected.

2 Study source E.
   a) Explain how intensive farming methods have reduced the number of skylarks.
   b) Is it possible for the countryside to be used for farming *and* conservation? Explain how.
   c) Using the skylark as an example, explain why managing the environment is an important part of conservation.

Homework

3 Do some research about one organisation involved in conservation. Find out what its aims are and what it does. The quickest way to do this would be on its website, using the internet. You can choose from this list, or find a local organisation.

- World Wide Fund for Nature (WWF) www.wwf-uk.org/
- The National Trust www.nationaltrust.org.uk
- Royal Society for the Protection of Birds (RSPB) www.rspb.co.uk
- British Trust for Conservation Volunteers (BTCV) www.btcv.org
- Greenpeace www.greenpeace.org
- Friends of the Earth www.foe.org.uk

Share the information with your class. You could choose one of these organisations to get involved with, either individually or as a class.

FRAMEWORK

4.4
Saving the planet

No one knows exactly how many plant and animal species live on our planet. Estimates range from five million to *30 million!* It is also estimated that between 3,000 and 30,000 of those species face **extinction** *each year*. Many more species are **endangered**, or will become extinct, if nothing is done to protect them.

One of the main threats to wildlife comes from the trade in endangered species. People trade anything from tropical fish for aquariums, to ivory from the tusks of African elephants. Most countries have signed an agreement – the Convention on International Trade in Endangered Species (CITES) – that makes trade in endangered species illegal. Even so, there is still a threat from illegal trade.

More important, perhaps, are the numbers of species endangered by the loss of their natural habitat. During the twentieth century, people destroyed the natural environment at an alarming rate. Half the world's rainforests were chopped down, and large areas of grassland, wetland and coral reefs disappeared. It is believed that thousands of species, most of them never identified, have been lost. Graph A shows the impact of human activity on the natural environment around the world.

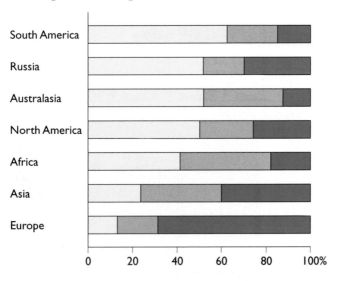

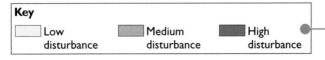

Key
Low disturbance   Medium disturbance   High disturbance

A The percentage of natural environment that has been disturbed

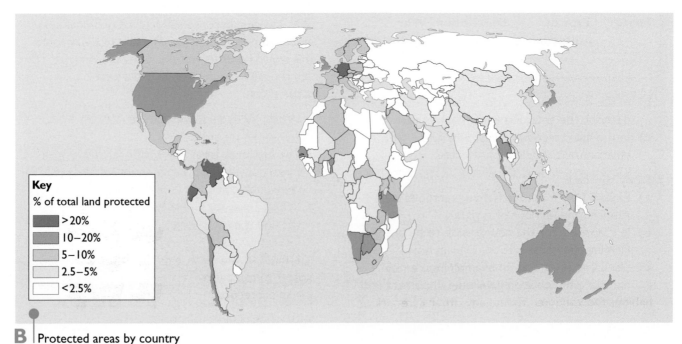

Key
% of total land protected
- >20%
- 10–20%
- 5–10%
- 2.5–5%
- <2.5%

B | Protected areas by country

5,000 left

The proboscis monkey lives in the swampy forests of Borneo. It is threatened by forest clearance and river pollution. It is also hunted for meat.

Only 500 left

The Chinese alligator lives in wetlands close to rivers. It is threatened by both floods and droughts, which happen more frequently in China these days.

8,000 left

The white rhino lives on the grasslands of central Africa. It is hunted for its horns which are worth more than their weight in gold.

2,000 left

The Caribbean manatee lives in tropical coastal water in the Caribbean Sea. It is threatened by accidental drowning when caught in fishing nets.

C

Activities

1 a) **N** Look at graph A. Which continent has:
   i) the largest areas of low disturbance?
   ii) the smallest areas of low disturbance?
   Give the percentage figures.

   b) How can you explain these differences?

2 a) Look at map B. Name five countries that have over 10 per cent of protected land. Check your answers in an atlas.

   b) Compare the percentage of protected land in the two continents from your answer to activity 1a. What do you notice? How can you explain the difference?

3 Look at the photos in C. Classify the endangered species into two groups: those endangered by illegal trade and those threatened by loss of habitat. Do you need to add any other categories?

You could carry out some further research to add to your lists. Try the website of the World Wide Fund for Nature at www.wwf-uk.org/

4 **L** The WWF uses the giant panda (an endangered species in China) on its logo. Not all endangered species are as appealing as the panda!

Choose one of the species in C. Think of all the reasons that is it important to protect a species like this. Draw a poster to campaign for its protection.

WWF ®

In this Building Block you will solve a mystery about why the English countryside is under threat. You will use the ideas to write a plan for one rural area.

4.5

What future for the countryside?

The English countryside is disappearing. Who is to blame? You are going to be geographical detectives to solve the mystery!

These families are two of the chief suspects! The Mason family in photo A have recently moved to the countryside. The Hall family in photo D have recently left their farm. What are their reasons for moving? And are they to blame for our disappearing countryside?

You will use different sorts of evidence to solve the mystery, starting with these photos and maps.

A The Mason family have moved to Bishop's Waltham in Hampshire.

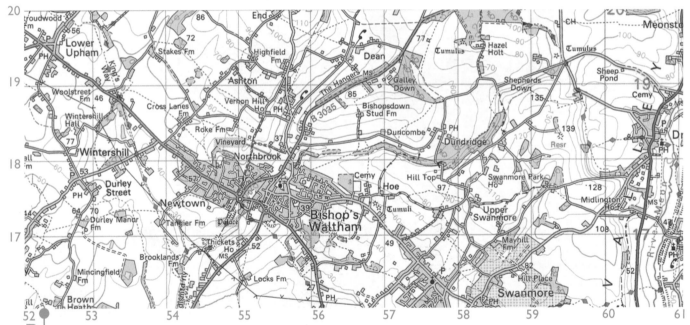

B 1:50,000 Ordnance Survey map extract of the area around Bishop's Waltham. Reproduced from the 2000 1:50,000 Ordnance Survey map by permission of the Controller of HMSO © Crown Copyright.

C Countryside in Hampshire, near Bishop's Waltham

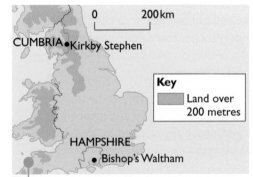

D The location of Bishop's Waltham and Kirkby Stephen

Activities

I Work with a partner. Think about the mystery that you have to solve. Devise a **theory** to explain why the two families are moving and how their decisions could be a threat to the countryside. Keep your theory, and as you work through this Building Block you can see if you are correct or whether you need to modify the theory.

E The Halls have left Stowgill Farm near Kirkby Stephen in Cumbria.

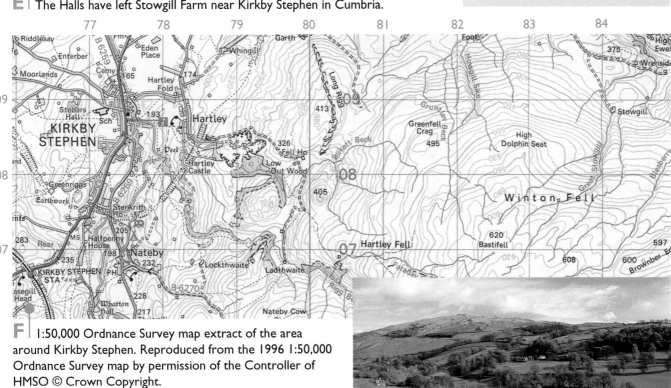

F 1:50,000 Ordnance Survey map extract of the area around Kirkby Stephen. Reproduced from the 1996 1:50,000 Ordnance Survey map by permission of the Controller of HMSO © Crown Copyright.

G Countryside in Cumbria, near Kirkby Stephen

2 With your partner, look at the maps and photos on these pages for evidence to solve the mystery. One of you can look in detail at map B and photo C. The other can look at map F and photo G.

What evidence do the maps and photos give you about:

- the physical environment
- settlement and population
- farming and other economic activities
- transport and services
- leisure and recreation?

Describe the evidence from the map and photo to your partner. Does the evidence help to explain why people are moving and how the countryside is under threat?

Together, decide whether the evidence helps to support your theory, or if you need to modify it.

How do rural problems differ from south to north?

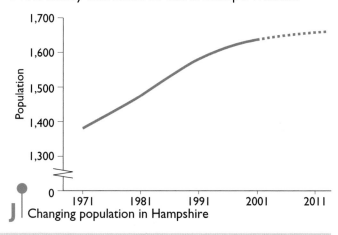

Bishop's Waltham
£295,000

Modern four-bedroom detached house with pleasant views to rear. The property is situated in the lovely village of Bishop's Waltham, with easy access to Southampton and surrounding towns.

Four bedrooms, one with en-suite bathroom • Luxury fitted kitchen/breakfast room with integral fridge and cooker • Hall • Lounge • Dining room with conservatory leading to garden • Downstairs cloakroom • Two bathrooms (including en-suite) • Garage and driveway • Gas central heating • Double-glazed windows and doors

I A recently-built house for sale in Bishop's Waltham

H Wickham Park Golf Club. One of many new golf clubs that have recently opened near Bishop's Waltham. Farmers can make more money from golf than they can from growing crops.

J Changing population in Hampshire

Activities

Continue to work with your partner to do these activities.

1 One of you should study the evidence in sources H, I and J (Hampshire) on this page and the other should study the evidence in sources K, L and M (Cumbria) on the next page.
   a) What does the evidence tell you about people moving to or from that area of the countryside?
   b) Suggest how this could be a threat to the countryside in that area.
   Together, decide if the evidence helps to support your theory, or whether you need to modify it.

2 There are different threats to the countryside in the two areas. The problems in Hampshire are different to those in Cumbria. Read the list of rural problems in source N.
   a) Decide whether each problem is more likely in Hampshire or in Cumbria. Make two lists, one for each area. Use all the evidence you have already found to help you.

   b) From your lists try to identify *one* key problem in each area. How does it link to all the other problems?
   c) Cut out copies of all the problems in one list. Your partner will do the same with the other list. Arrange them on a sheet of paper with your key problem at the centre. Stick the problems down. Draw an arrow to link each problem with the key problem. Write a sentence on each arrow to explain the link. You can use the diagram you have made when you do the assignment on page 80.

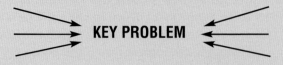

KEY PROBLEM

3 Do you want to make any final modifications to your theory? The solution to the mystery is revealed on page 80. Check to see how close you came to the truth.

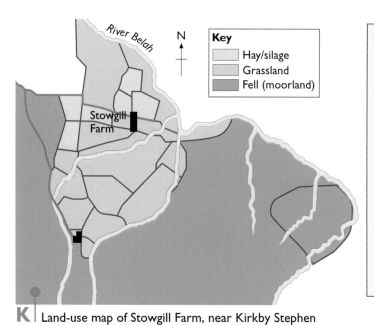

Key
- Hay/silage
- Grassland
- Fell (moorland)

Stowgill Farm

River Belah

N

Factfile
Location: Kaber, near Kirkby Stephen. Grid reference: 842 088 on map F
Owners: Maurice and Linda Hall
Relief: hilly land with valley of River Belah to north and open moorland to west
Altitude: 360 m average
Soil: peat over limestone, poor fertility
Average annual rainfall: 1,250 mm
Land use: grassland 68 ha, rough grazing 12 ha, common fell grazing 400 ha
Livestock: 1,440 sheep
Labour: two full-time family members
Income: 40 % of income from sale of lambs and wool
          60 % of income from government subsidies

K Land-use map of Stowgill Farm, near Kirkby Stephen

L Cumbria was one of the worst affected areas in the foot-and-mouth outbreak of 2001: 750 farms in Cumbria were affected and over 700,000 sheep were slaughtered.

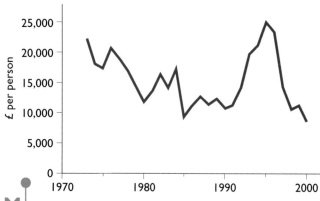

M Average UK farm income. Farmers in Cumbria earn below the national average.

There is too much traffic in the village.	Population is growing fast.	Farmers use intensive methods that harm wildlife.
There is only one shop and a pub in the village.	It is 40 kilometres to the nearest supermarket.	Holiday homes are empty for most of the year.
There is no money in livestock farming.	Young people can't afford to buy homes.	There is no public transport.
Local schools are over-subscribed (more pupils than places).	There is a shortage of land.	Population is falling.
Houses take up land.	The moors are becoming overgrown.	Local schools have closed.
There are few jobs apart from farming.	Land is very expensive.	Stone walls are falling into disrepair.
Golf clubs have taken over good farmland.	Farmers depend on government subsidies (financial help).	The soil is poor and land is steep.
Meadows and hedges are being ploughed up.	Young people want to leave.	

N Rural problems

BUILDING BLOCKS

How would you plan the countryside?

Finally, the solution to the mystery can be revealed. Here both families tell you about their decisions to move.

○ Andy Mason explains why they decided to move to Bishop's Waltham.

> We moved to Hampshire from London. We had thought about moving to the countryside for years. We were fed up with all the hustle and bustle and wanted a better environment for the kids. I was lucky enough to get a new job in Portsmouth. We were able to move to a village that was within easy driving distance.

> We bought Stowgill Farm fifteen years ago. The work was hard, we didn't make much money, but the life was good. Now we're getting older and our children don't want to take the farm over. Out of the blue we were made an offer by a company that wants to use the land for grouse-shooting. We were lucky to sell it before the foot-and-mouth outbreak began. We could have been bankrupt by now.

P| Linda Hall explains why they decided to move from Stowgill Farm.

The Masons and the Halls are typical of families in two areas of England, where many are making similar decisions. So how could their decisions be a threat to the rural environment?

On the one hand, people are moving to the countryside from cities for a better environment and quality of life. As a result the countryside is becoming urbanised. This is particularly true in the more densely-populated south of England. On the other hand, as farm incomes fall, farmers find it harder to make a living and are being forced to leave their land. As a result the landscape is changing and villages are dying. This is particularly true in upland areas, such as the north of England. In both cases, the English countryside that we know could disappear.

Assignment

You are going to prepare a plan for a rural area. You can choose one of the two areas that you have studied in this Building Block: the area around Bishop's Waltham in Hampshire, or the area around Kirkby Stephen in Cumbria. If you live in the countryside you could prepare a plan for your own area. Which of the two areas is it more like?

First, you need to think about the sort of countryside that you want in future. Do you want to keep the rural environment the way that it has always been, or do you want it to change? Do you want to see more people living there, or fewer? Will farming still be an important part of the countryside?

Next, look at the box on the right. It has some strategies that you could use in your plan. Choose up to five strategies that you think would work in your area or add your own.

Write your plan. Write a section to explain each of the strategies that you have decided to use. The menu on the opposite page will help you.

- Pay farmers to look after the environment, not to produce food.
- Build more houses.
- Restrict activities that use up farmland.
- Abandon farmland and let it grow wild.
- Build factories and offices to provide rural employment.
- Encourage farmers to switch to tourism to earn a living.
- Don't allow any more houses to be built.
- Improve cities so that fewer people would want to leave.
- Encourage farmers to use less intensive methods.
- Provide better public transport.

How to write a plan

Text type	Tense	Starters	Links	Conclusions	Vocabulary
Recount	Past	First, …	… and …	In conclusion, …	countryside
Description	*Present*	Second, …	… also …	In summary, …	rural
Method	*Future*	In the first place, …	… as well as …	Overall, …	environment
Explanation		The main priority …	… so …	On the whole, …	urbanisation
Persuasion		Later …	… because …	I suggest that …	farming
Discussion			Above all, …	I recommend …	conservation
			… due to …	Finally, …	intensive farming
			As a result of …		habitat
			For example, …		housing
			… such as …		recreation
			… mainly …		
			… usually …		
			Unfortunately …		

STRUCTURE FOR WRITING A PLAN

1 The title should say where your plan is for. For example, *Plan for the area around Bishop's Waltham.*

2 Write an introductory paragraph about the area to provide a background to your plan. Describe its location and the main physical and human features, including relief, climate, population and settlement. Outline the main problems or issues that your plan will deal with. You could use a map to help to describe the area and to highlight the main strategies in your plan.

3 Divide your plan into sections. Each section should deal with one of the main strategies in your plan. Think of a snappy title for each strategy. For example, *Don't allow any more houses to be built* could be titled *No more houses.*

   Write a first paragraph to explain what the problem or issue is that you want to solve. This should be written in the third person and present tense. For example, *Large areas of countryside are being taken up with new houses* … The second paragraph will explain how your strategy will help to solve the problem and could be written in the first person and future tense. For example, *In the first place I/we will limit the number of new houses that are built* …

4 Write a final paragraph to sum up the main points in your plan. Explain what the overall impact of your plan would be on the area.

BUILDING BLOCKS

In this Building Block you will investigate Antarctica. Could human activities there be a threat to the natural environment? You will decide.

4.6

Should the penguins have a vote?

To	Account	Date Sent

To: St Edward's C of E Middle School, Leek, Staffs
From: Steve Marshall, Evans Ice Stream, Antarctica, 76 38'S 79 12'W

Hello all,
I hope you're OK back there in the UK and not all catching the flu we've been hearing about. Not much chance of it reaching us out here 20,000 kilometres away!

Well, I've been out here on the Evans Ice Stream for the past six weeks along with seven colleagues trying to make a seismic survey of the area. I say 'trying' as we are having a hard time with the awful weather. For half the time out here I've been stuck in my tent, with blizzards raging outside. It's in one of these blizzards now that I write to you! The temperature outside is 20 degrees below freezing, the wind is 20 knots and you can see no more than a few metres because of the blinding snow. It would be easy to lose your tent if you went outside. In these conditions that would mean certain death! So when it's like this we don't go outside unless it's absolutely necessary.

On a nice day though this place is awesome. A white desert rolling to every horizon with an unimaginable vastness. The Sun at this time of year does not set and just circles around our heads, only dipping slightly towards the southern horizon to tell us it's night-time!!

I suppose this place is as close to being on the moon as you can get on Earth.

A Steve Marshall is working with the British Antarctic Survey. While he was in Antarctica he kept in touch with pupils at a school in the UK by e-mail.

Antarctica is the world's last great **wilderness**. It is a frozen continent that is almost entirely buried beneath snow and ice. No people live here permanently – only a few scientists from various countries based at scientific stations dotted hundreds of kilometres apart (map C). Why would anybody be interested in such a remote place?

B

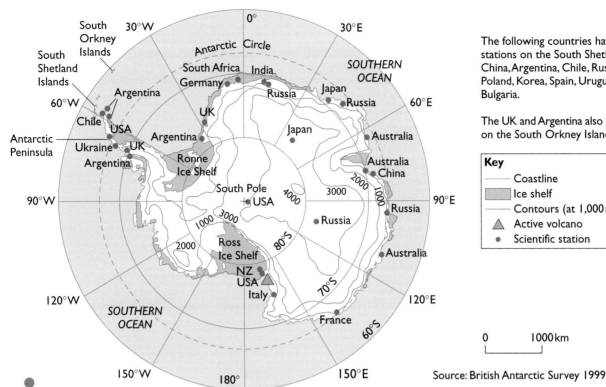

The following countries have scientific stations on the South Shetland Islands: China, Argentina, Chile, Russia, Brazil, Poland, Korea, Spain, Uruguay and Bulgaria.

The UK and Argentina also have stations on the South Orkney Islands.

Key

——	Coastline
▨	Ice shelf
——	Contours (at 1,000 m intervals)
△	Active volcano
•	Scientific station

0   1000 km

Source: British Antarctic Survey 1999

C Antarctica

Activities

1 [L] Work with a partner.

One of you should read e-mail A aloud while the other person looks at photo B. The person that is looking should imagine that they are inside the tent during a blizzard.

- What can you see?
- What can you hear?
- What can you smell?
- How do you feel?

Now reverse roles. This time the person looking has to imagine that they are outside the tent on a clear day. Answer the same four questions.

2 Now put your geographer's hat on and read e-mail A again carefully.
  a) What useful geographical information does it give you about Antarctica? Make notes. You will do this again when you come to further extracts from Steve's e-mail later.
  b) What questions would you like to ask Steve? These could be things you don't understand, or things you'd like to know more about. Try to find your own answers in the rest of this Building Block.

D Rothera Scientific Station: one of the stations where the British Antarctic Survey is based

  c) Use the grid reference in e-mail A to locate Steve's position on map C.

3 Look at map C and photo D.
  a) List the countries with scientific stations in Antarctica.
  b) Describe the distribution of the stations. How could you explain this distribution?
  c) Suggest why countries might want to have scientific stations in Antarctica.

Why is anybody here?

To	Account	Date Sent	!	📎	Subject

What are we doing here? You may well ask! Not easy to explain this, but here goes. A seismic survey is looking to see – or should I say hear – what's under our feet. I'm sitting on ice which is thousands of metres thick and underneath that there's rock. The scientists are interested in the thickness of the ice and the mineral-bearing rock.

Imagine going into a room with your eyes closed. You don't know the size of the room and so you shout. From what you hear you get some idea of the size of the room. If it were a small cubicle your shout would sound loud and short, but if you were in a large hall the same shout would sound quieter and there'd be an echo.

Here, instead of shouting at the snow, we use 500 kg of high explosive to make the noise. Then we record the sound echo with scientific equipment. From the data we can discover the thickness of the ice, the type of rock below and even what it contains.

E Steve's e-mail

F Exposed rock in Antarctica. Can you spot a layer of coal? It is possible that larger mineral deposits lie beneath the ice. If a way is found to extract minerals like coal and oil from below the ice, we could use them if we need extra resources in the future.

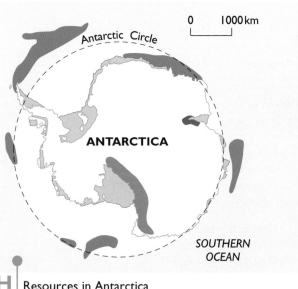

H Resources in Antarctica

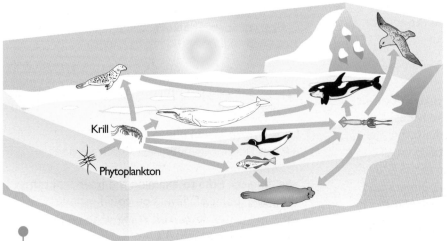

G The ecosystem of the Southern Ocean

The Southern Ocean contains a surprising number of species given the harsh conditions. This is only possible because of the abundance of phytoplankton (microscopic plants that trap the Sun's energy) and krill (small, shrimp-like creatures that eat the phytoplankton). Fish, penguins, seals, whales and seabirds, common around the Antarctic coast, all depend heavily on krill in their diet. Commercial fishing for krill and fish was becoming a threat to the whole **ecosystem** of the Southern Ocean. This is now controlled by quotas on the amount that can be caught. However, illegal fishing is still a problem.

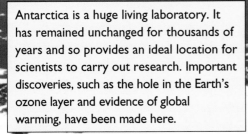

Antarctica is a huge living laboratory. It has remained unchanged for thousands of years and so provides an ideal location for scientists to carry out research. Important discoveries, such as the hole in the Earth's ozone layer and evidence of global warming, have been made here.

Early research stations did not take much care of the environment. Sewage went into the sea and rubbish was dumped. Now scientific stations are a model for us all – sewage is treated before it goes into the sea and rubbish is removed from Antarctica.

Most research stations are built on ice-free sites around the edge of Antarctica. These ice-free zones also support most of the continent's plant and animal species, so people are in direct competition with wildlife.

Construction of airstrips can involve levelling of land and destruction of the shoreline. It also creates noise and pollution that disturb the environment over a much wider area.

I Scientific stations and the environment

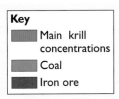

Key
- Main krill concentrations
- Coal
- Iron ore

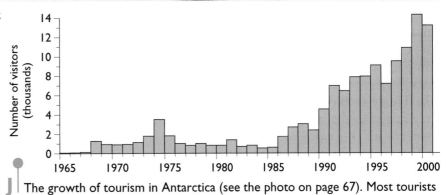

J The growth of tourism in Antarctica (see the photo on page 67). Most tourists arrive by ship and spend little time on land.

Activities

1 Read extract E from Steve's e-mail.
  a) Put your geographer's hat on again and note any new geographical information.
  b) What further questions would you like to ask? Add them to your list.
  c) Does the other information on pages 84–85 answer any of your questions?

2 Study all the information on these two pages. Draw a large table with three columns. In the first column, list four possible human activities in Antarctica: fishing, scientific research, tourism and mining.

In the second column, explain why Antarctica is, or could become, an important place for these activities.

In the third column, suggest what impact each activity could have on the environment.

3 Now you are going to do something that is not very scientific! Imagine that you are a penguin living in Antarctica. You and your ancestors have lived here for thousands of years. But recently a new species has arrived – humans!

Penguins are divided. Some see the new arrivals as a threat, both to penguins and the environment as a whole. Others see them as an opportunity for long overdue development in Antarctica.

Study all the evidence on pages 82 to 85. What do you think? Are humans a threat or an opportunity for Antarctica? Write an article for the *Antarctic Echo* to examine the issue from the penguins' point of view. You could include imaginary interviews with some of the new arrivals to get a human angle!

It is not just scientist who are interested in Antartica – governments and big companies are interested too. You are going to play a game about future development of Antartica and write a new treaty on how it could be governed.

4.7

Can Antarctica be governed globally?

The Antarctic Treaty, agreed in 1961, established Antarctica as a region of peace and science. Now 45 countries, representing over three-quarters of the world's population, have signed the Treaty. It bans all military activity in Antarctica and encourages scientific research in Antarctica and co-operation between the countries that work there. Unlike other international treaties every decision is made by consensus (reaching agreement) not by majority vote.

The Treaty was put to the test in the 1980s when mining companies wanted to search for minerals in Antarctica. Environmental groups like Greenpeace strongly opposed this and campaigned for the continent to become a World Park, where mining would be banned for ever. The Treaty countries failed to agree but eventually, they negotiated a new Protocol on Environmental Protection. This came into effect in 1998.

The Protocol sets out rules about the conservation of wildlife, protection of habitats and control of pollution. It also puts a 50-year ban on any mining in Antarctica unless or until all the Treaty countries agree to allow it. At present, cheaper and more accessible mineral resources can be found elsewhere.

Do you think this Protocol will last when other sources of minerals are no longer available? What would happen if valuable minerals, like gold or oil, were discovered in Antarctica? Is it possible for 45 countries to agree? You are going to play a game to find out.

Activity

Play the game with a partner, or in a group of three. You represent one of the countries with a scientific station in Antarctica. The rest of your class will represent the other countries. Your teacher will tell you which country to be. As the game proceeds you will be faced with decisions to make: should you *compete* or *co-operate*? Remember, the Antarctic

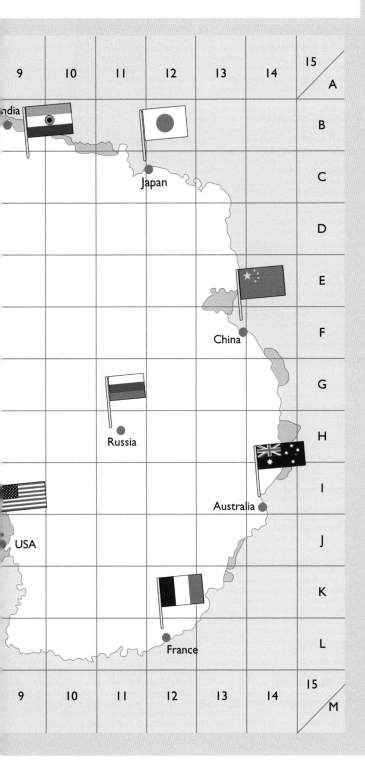

Treaty encourages countries to work together. However, if you discover a valuable mineral it could make your country much richer. Decisions, decisions . . .

How to play

1 First, place a token on the square where your scientific station is. You are going to explore Antarctica. Take turns, with all the other groups in the class, to move your token. (You could do this in alphabetical order, or some other way.) On each turn, you can move your token one square in any direction: north, south, east or west. You are allowed to explore anywhere on the map, on land or sea.

2 Every time you land on a square, carry out a seismic survey (explained in extract E on page 84) to find out if there are any minerals there. (Your teacher will tell you what is there. They should have a secret map that locates the mineral deposits in each square.)

3 After the first round, when every country has had a turn, confer with your partner or group. You must decide whether to compete or co-operate with the other countries.

- If you decide to *co-operate*, vote to continue the Environmental Protocol. Remember, mining can only happen if every country agrees to allow it.
- If you decide to *compete*, vote to end the Environmental Protocol, then you will be able to extract the minerals that you have found. However, you will not be allowed to mine unless or until *all* the countries agree to end the Protocol.

4 Play the next round of the game. Each group continues to explore for minerals. You should keep a record of all the minerals you find, which round you find them in, the squares they are in and what they are worth. Later, if there is any dispute about who the minerals belong to, you can prove that you found them first.

   At the end of each round, again confer before you vote whether to continue the Environmental Protocol. Play the game until the whole continent has been explored, or you teacher tells you it's time to stop.

5 What was the outcome of the game? Did you agree to continue the Environmental Protocol, or to end it? Why did this happen, do you think? Is the same thing likely to happen in real life?

A *new Antartic Treaty*

Assignment

You have to devise a new Antarctic Treaty that is in the best interests of the environment and all the people of the world.

There are two main views about the future of Antarctica.

The conservationist viewpoint is put forward by Greenpeace. They argue that Antarctica should become a World Park – like a national park on a bigger scale. It could be governed by the United Nations organisation on behalf of all the countries in the world. Activities that harm the environment would be totally banned. This would preserve Antarctica as the world's last great wilderness and be in the best interest of the continent's wildlife, which does not have a vote of its own.

The other view is that Antarctica is already the most protected place on Earth. It should continue to be governed as it is now – by the countries that have signed the Antarctic Treaty. It has proved that the environment can be protected and that it is possible to stop harmful activities such as mining.

Write a treaty that addresses the following questions:

- Who should govern Antarctica?
- Which countries should be represented?
- How would decisions be taken?
- How would the environment be protected?
- Which activities would be permitted?
- Which activities would be banned?
- How would the treaty be enforced?

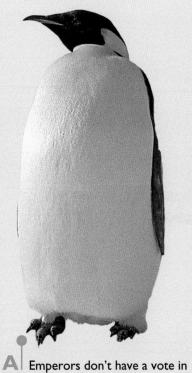

A Emperors don't have a vote in Antarctica!

To	Account	Date Sent	!	📎	Subject

Anyway, back here in my tent, things are not that bad. I know it's –20 °C outside, but in here it's cosy and warm. The Tilly Lamp is on and the CD player is drowning out the sound of the wind outside. The most horrendous experience here is going to the toilet. All I can say about this is it's done very quickly!!!

Two days ago, Alfie (my tent partner) and I travelled by skidoo to our supply depot. We covered over 150 kilometres without seeing any change in the landscape at all. We had to follow a compass bearing, just as if we were crossing an ocean. Eight hours later, when we arrived at our destination, it looked no different from the place we had left – flat and white!!

I had better close there. It's time to start cooking some dinner. I hope this letter gives you some idea of my life in the freezer. It's a hard place to live, but also very rewarding. Look forward to hearing from you all.

Regards,
Steve Marshall

B

Extra

 You may still have questions about Antarctica. Try to e-mail scientists working there. Find out who you can write to by contacting the British Antarctic Survey at www.antarctica.ac.uk

UNIT 5   CLIMATE CHANGE –
When is a disaster no longer natural?

People say that weather around the world is becoming less predictable. Extreme types of weather are becoming more common. Geographers describe the weather events that you can see in the photos as **natural hazards**. But how natural are they?

- What types of weather can you see in these photos?
- Have you ever experienced any of them? Where were you at the time?
- In what ways could each of these events be a hazard? What would make them a disaster?
- Can you think how people might be responsible for causing any of these events?

GROUNDWORK

5.1 Was the sinking of the Titanic *a natural disaster?*

Did you see the movie *Titanic*? It is the true story of the 'unsinkable' ship that sank in 1912 after it hit an iceberg in the Atlantic Ocean. It was the worst ever disaster at sea. More than 1,500 passengers and crew died. But can this be described as a *natural* disaster, or were people to blame? Read the full story to find out.

A The *Titanic* at the start of its maiden voyage

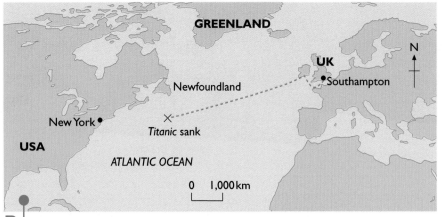

B The route taken by the *Titanic* across the Atlantic Ocean

Activities

1 Read the story of the *Titanic* on these two pages. Identify as many reasons for the disaster as you can. How many of these were natural? How could people be blamed?

2 Was the sinking of the *Titanic* a natural disaster? Discuss this question with a partner.

When the *Titanic* steamed out of Southampton, bound for New York, on its maiden voyage on 10 April 1912, it was the largest liner that had ever sailed. Built at a cost of one million pounds (nearer £100 million in today's money), it weighed 46,000 tons and was the most up-to-date ship of its type. Its hull was double-bottomed and divided into sixteen watertight compartments. Even if four of these compartments were to flood the liner would still remain afloat. One employee of the White Star Line, which owned the *Titanic*, boasted, 'God Himself could not sink this ship!' The owners were so confident that they fitted just twenty lifeboats – enough to hold just half of the 2,200 passengers and crew on board.

The *Titanic* was the last word in luxury. First-class accommodation included four-poster beds, oriental carpets and crystal chandeliers. Many of the passengers on the maiden (first) voyage were rich and famous. However, the *Titanic* also offered cheaper accommodation, known as steerage, for third class passengers sharing bunk beds in large dormitories on the lower decks. Many of them were poor migrants from Europe on a one-way ticket, heading for a new life in America.

Four days into the voyage, the *Titanic* was speeding through the North Atlantic, south of Newfoundland. The ship's owners were aiming to break the record for the fastest Atlantic crossing. Despite earlier warnings of icebergs in the region, the ship carried on full-steam ahead. At 11.40 p.m. on Sunday 14 April the ship's look-out was high up in the crow's nest on the forward mast. The binoculars were lost and he was peering into the darkness (hi-tech equipment such as radar, sonar and global satellite positioning had not yet been invented). He could faintly see a large object, growing in size as they drew closer. 'My God!', he exclaimed as he grabbed the pull cord on the ship's bell. 'Iceberg ahead!' he informed the officer on the bridge below.

An iceberg is a floating mass of ice that has broken from the end of a glacier where it reaches the sea. Most of the iceberg lies below the sea surface so that only about one-tenth is visible from above the water (hence the saying, 'The tip of an iceberg'). They form mostly in spring or summer, when warmer weather increases the rate of calving (separating) from the main glacier.

Icebergs are found in both the Arctic and Antarctic. Arctic icebergs vary in size from a large piano to a ten-storey building. About 10,000 icebergs are produced every year from Greenland's glaciers, some of which flow south of Newfoundland into the North Atlantic where they can be a hazard to shipping.

C   The sinking of the ship in the film *Titanic*

As soon as danger was spotted the officer tried to turn the *Titanic* and put the engines into reverse. There were a few tense minutes as the ship slowly responded, and relief when a head-on collision was averted. But then there was a slight shudder as the iceberg scraped along the starboard side of the boat. The passengers barely noticed and the iceberg quickly passed by. But inspection revealed that five of the watertight compartments in the hull were ruptured and had begun to let in water.

Captain Smith gave the order to 'Abandon ship!' At first the passengers thought it was silly to get into the lifeboats as the *Titanic* was supposed to be unsinkable! The first few lifeboats were lowered with only a handful of people in them. Soon water began to fill the lower decks and the ship began to dip forward. Now there was a rush to get on deck and the steerage passengers were held back to allow first-class passengers to reach the lifeboats. Among the first passengers to escape the sinking liner was the managing director of the White Star Line!

The last few lifeboats were filled to capacity and the rule was 'Women and children first!' Survivors watched the ship slide into the water as its lights flickered out.

One survivor, wireless operator Harold Bride, had to swim for his life. 'The ship was gradually tilting onto her nose – just like a duck that goes for a dive. I had only one thing on my mind – to get away from the suction. The band was still playing,' he recalled later. 'I guess all of them went down.'

Less than three hours after it hit the iceberg, the *Titanic* sank. Of the 2,200 people on board just 703 survived. Captain Smith and most of his crew went down with the ship.

Assignment

As you study the rest of this unit keep the story of the *Titanic* in the back of your mind. Through the next twenty pages you will investigate global **climate** change, floods in the UK and drought in Africa. What evidence can you find that these are natural disasters or that people are to blame?

Keep a record of the evidence that you find under two headings:

Evidence that disasters are natural

Evidence that people are to blame

You will use the evidence that you collect in the final assignment at the end of the unit.

FRAMEWORK

5.2 **W**eird weather

People in the UK talk about the **weather** a lot. It is a national joke (cartoon A). But now there's really something to talk about. The weather seems to be changing.

Do you remember the winter of 2000–1? The headlines in B might jog your memory. It was the wettest winter in this country since records began. Some parts of the UK had double their average rainfall. This was just one of a series of unusual weather events in recent years. During the 1990s the UK had the longest drought in living memory and throughout the decade temperatures were getting hotter. Seven of the hottest years ever recorded were in the 1990s, and 1999 was the hottest of all.

There's nothing weird about the weather – it's always raining cats and dogs!

A | The weather can be no joke

We're all in the same boat

Global warming – it's with us now

Freak tornado batters Bognor

Six dead in storm chaos

Floods bring more misery

B | Headlines from the winter of 2000–01

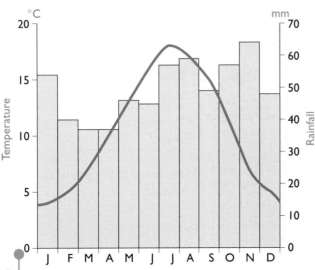

C | Average monthly temperature and rainfall in London recorded by the Met Office

- **Weather** is day-to-day changes in the atmosphere.
- **Climate** is the average weather pattern recorded over many years.

To prove climate is changing, we need to keep accurate weather records for many years. This is what the **Meteorological Office** has been doing in the UK for over 100 years (see graph C).

Start your investigation about disasters on these two pages. Remember, you are looking for evidence that disasters are natural or that people are to blame (see page 91).

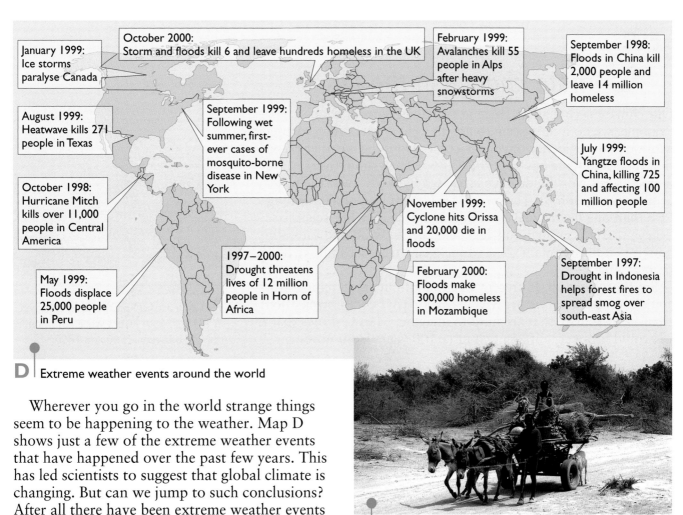

January 1999: Ice storms paralyse Canada

October 2000: Storm and floods kill 6 and leave hundreds homeless in the UK

February 1999: Avalanches kill 55 people in Alps after heavy snowstorms

September 1998: Floods in China kill 2,000 people and leave 14 million homeless

August 1999: Heatwave kills 271 people in Texas

September 1999: Following wet summer, first-ever cases of mosquito-borne disease in New York

July 1999: Yangtze floods in China, killing 725 and affecting 100 million people

October 1998: Hurricane Mitch kills over 11,000 people in Central America

November 1999: Cyclone hits Orissa and 20,000 die in floods

May 1999: Floods displace 25,000 people in Peru

1997–2000: Drought threatens lives of 12 million people in Horn of Africa

February 2000: Floods make 300,000 homeless in Mozambique

September 1997: Drought in Indonesia helps forest fires to spread smog over south-east Asia

D Extreme weather events around the world

Wherever you go in the world strange things seem to be happening to the weather. Map D shows just a few of the extreme weather events that have happened over the past few years. This has led scientists to suggest that global climate is changing. But can we jump to such conclusions? After all there have been extreme weather events since the time of Noah!

E Drought affects millions in Africa

Activities

I Look at the headlines and the photo in B.
   a) Can you remember the winter of 2000–01? Describe any weather events you can recall and the effect they had.
   b) Describe any other unusual weather events that you can remember. When, and where, did they happen?

2 a) Observe and record temperature and rainfall in your area each day for a week. Do this at the same time every day at your school weather station. Your teacher will give you instructions on how to do this.
   b) Draw graphs to show the daily temperature and rainfall you recorded during the week. Use a line graph to show temperature and a bar graph for rainfall, as in graph C. Work out the average daily temperature and rainfall. What would the monthly rainfall be?

   c) How do your averages compare with the monthly averages in graph C? If you want to find the average temperature and rainfall in your area, you can use the Met Office website at www.metoffice.com/education.
   d) Do your records provide any evidence for climate change? How reliable is the evidence? Explain your answers.

3 Look at map D.
   a) List the different types of weather event mentioned on the map. What other hazards did some of these events lead to?
   b) Which hazards appear to cause the greatest disasters? Give reasons for your answer. Think about the loss of life, loss of homes and land, the area affected and the duration.
   c) Where did the worst disasters occur? Can you suggest why?

5.3 Climate change in the UK

Our climate *is* changing. There are obvious signs of change. For example, we no longer seem to get the cold winters that we used to. More reliable evidence of change comes from the weather records that the Met Office has kept since 1860. They show that temperatures worldwide have increased by 0.6°C over the past century (see graph B). This process is known as **global warming**. Less than one degree may seem like a tiny change but scientists believe that it is enough to trigger changes in wind and rainfall patterns. It may explain some of the weird weather that we have been having.

But that is not all. If present trends continue, then temperature could increase by a further 3°C by 2100. That could bring about changes in climate that are hard to imagine. Map C shows that the UK could be a very different country 100 years from now.

A A 'frost fair' on the River Thames in 1814. Frozen rivers were common in the UK between 1500 and 1850

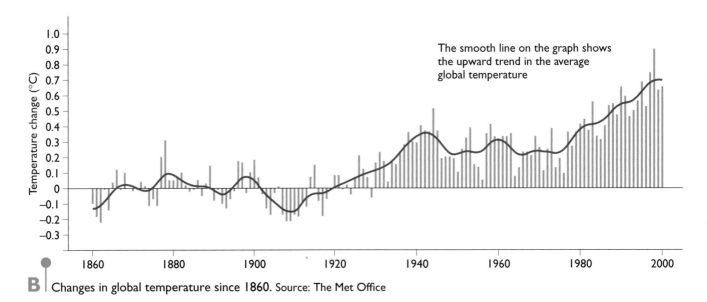

The smooth line on the graph shows the upward trend in the average global temperature

B Changes in global temperature since 1860. Source: The Met Office

Activities

1 Look at graph B.

a) List the five coldest and the five warmest years.

b) Describe carefully how global temperature has changed over the past 140 years. Use the following words in your description:

> trend average fluctuation increase
> exceptional global warming

You can use a dictionary to help you.

2 Look at map C. From the evidence on the map, how do you think that climate change will affect the UK?

a) Make a table with two columns:

- positive changes
- negative changes.

List each of the changes on the map under one of the two headings.

b) Should we be worried about climate change in the UK? Give reasons.

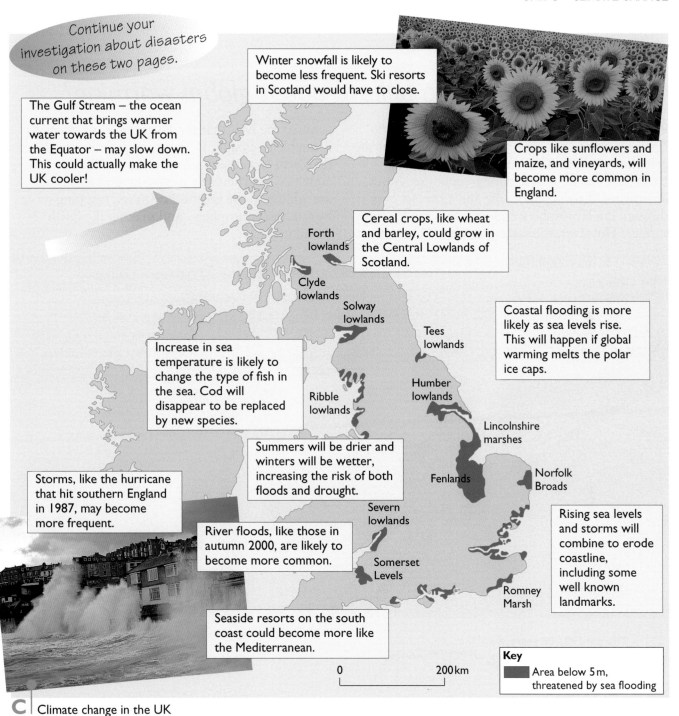

Continue your investigation about disasters on these two pages.

Winter snowfall is likely to become less frequent. Ski resorts in Scotland would have to close.

The Gulf Stream – the ocean current that brings warmer water towards the UK from the Equator – may slow down. This could actually make the UK cooler!

Crops like sunflowers and maize, and vineyards, will become more common in England.

Cereal crops, like wheat and barley, could grow in the Central Lowlands of Scotland.

Coastal flooding is more likely as sea levels rise. This will happen if global warming melts the polar ice caps.

Increase in sea temperature is likely to change the type of fish in the sea. Cod will disappear to be replaced by new species.

Summers will be drier and winters will be wetter, increasing the risk of both floods and drought.

Storms, like the hurricane that hit southern England in 1987, may become more frequent.

River floods, like those in autumn 2000, are likely to become more common.

Rising sea levels and storms will combine to erode coastline, including some well known landmarks.

Seaside resorts on the south coast could become more like the Mediterranean.

Forth lowlands

Clyde lowlands

Solway lowlands

Tees lowlands

Humber lowlands

Ribble lowlands

Lincolnshire marshes

Fenlands

Norfolk Broads

Severn lowlands

Somerset Levels

Romney Marsh

0   200 km

Key
Area below 5 m, threatened by sea flooding

C | Climate change in the UK

3 Suggest how each of the following people is likely to be affected by climate change in the UK:

- a fisherman in the North Sea
- a farmer in south-east England
- a hotel owner in Devon
- a resident of York close to the River Ouse
- the owner of a ski resort in Scotland.

4 The UK has an important tourist industry. How will tourism differ by 2100?

Design a tourist brochure to attract foreign tourists to the UK in 2100. Think about the new tourist attractions that will have grown by then, as well as the ones that may be lost. Include a new tourist map of the UK in your brochure. You can use a desktop publishing package to make your brochure look good.

5.4 What should we believe about global warming?

Predictions about global warming in the twenty-first century are based on what we know about the **greenhouse effect**. The atmosphere around the Earth acts like glass in a giant greenhouse. Gases, like carbon dioxide, allow the Sun's rays to pass through the atmosphere and warm up the Earth's surface. However, the same gases also help to trap some of the heat that the Earth gives off, so that it is unable to escape. Without the greenhouse effect the Earth would either boil or freeze. But human activities – particularly the burning of **fossil fuels** – are adding to the greenhouse gases, trapping more heat in the atmosphere, and making the Earth warm up.

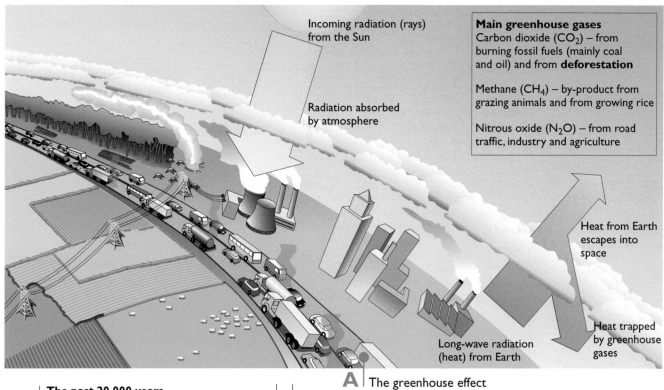

Incoming radiation (rays) from the Sun

Radiation absorbed by atmosphere

Main greenhouse gases
Carbon dioxide (CO_2) – from burning fossil fuels (mainly coal and oil) and from **deforestation**

Methane (CH_4) – by-product from grazing animals and from growing rice

Nitrous oxide (N_2O) – from road traffic, industry and agriculture

Heat from Earth escapes into space

Long-wave radiation (heat) from Earth

Heat trapped by greenhouse gases

A The greenhouse effect

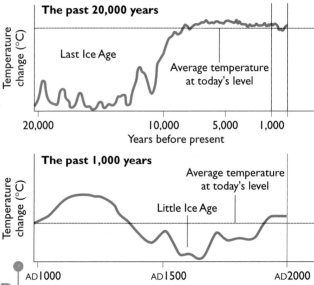

The past 20,000 years

Temperature change (°C)

Last Ice Age

Average temperature at today's level

20,000   10,000   5,000   1,000
Years before present

The past 1,000 years

Temperature change (°C)

Average temperature at today's level

Little Ice Age

AD 1000   AD 1500   AD 2000

B Changes in global temperature

Global temperature has been increasing since records began (see graph B on page 94). But, if we go back further, we find that temperatures have been going up *and* down (source B on this page). Scientists examined ice cores from the North and South Poles to investigate climate in the past. What they discovered is that the Earth has been through a number of Ice Ages. The last one ended 10,000 years ago. Even during the past 1,000 years there have been smaller fluctuations in temperature. During the Little Ice Age, 1500–1850 in Europe, rivers froze in winter. So, global warming could be part of one of these natural cycles.

Continue your investigation about disasters on these two pages.

C Predicted temperature change from now until 2080, assuming that we continue to burn fossil fuels at the present rate.
Source: The Met Office

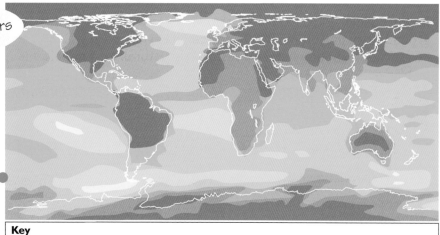

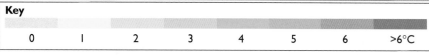

Key								
0	1	2	3	4	5	6	>6°C

Scientists use complex computer models to predict how global warming will affect temperature and **precipitation** in different parts of the world (maps C and D). These may look very convincing, but the problem with such models is that they are only as good as the information that was fed into the computer in the first place! And though scientists have improved their understanding of climate, it is still far from perfect.

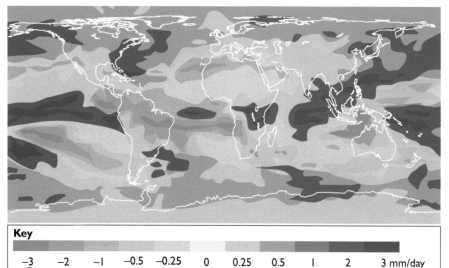

Key										
−3	−2	−1	−0.5	−0.25	0	0.25	0.5	1	2	3 mm/day

D Predicted change in annual precipitation from now until 2080, based on the temperature changes in map C

Activities

1 Look at drawing A.
   a) Identify five sources of greenhouse gases in the drawing. In each case, say which gas or gases are being produced.
   b) Explain how the increase in greenhouse gases could lead to a rise in temperature.

2 Look at the graphs in B.
   a) Describe changes in global temperatures over:
      i) the past 20,000 years
      ii) the past 1,000 years.
   b) Compare these graphs with graph B on page 94. Where would that graph fit onto the two graphs here? What does this suggest about the changes in global temperature over the past 140 years?

3 Look at maps C and D.
   a) Identify six parts of the world that are expected to have the largest rise in temperature (6°C or more) by 2080.
   b) Identify three parts of the world that are expected to have:
      i) the largest rise in precipitation (3 mm a day)
      ii) the largest fall in precipitation (3 mm a day)
      by 2080.
   c) What changes do the maps suggest we are likely to experience in the UK?

4 Does the evidence on these pages help to prove, or disprove, the idea of global warming? Give at least two reasons to support your opinion.

In this Building Block you are assigned to report on the floods around York in Autumn 2000. Include this evidence in your investigation about whether disasters are natural, or if people are to blame.

5.5

W*ill we have to get used to wet weather?*

It's been raining again . . . heavily! Many parts of Britain are under water. Roads are closed and railway services are suspended. The only way to travel in these areas is by helicopter. Does this sound like make-believe? Well, it happened in the autumn of 2000. If scientists' predictions about global warming are correct, then flooding will become a more frequent hazard in the UK.

A Flooding on the River Ouse near York in November 2000

Assignment

You can do this assignment on your own or work with a partner.

You are reporting for a national newspaper. You have been assigned to write a 500-word report about the floods affecting the area around York. Your first problem is how to get there. The city is virtually cut off by floodwater. You have to take a helicopter.

As the helicopter circles the city, looking for a dry landing site, you are greeted by the view in photo A. You are shocked. Where did all this water come from? Why wasn't the city protected against flooding? What's happened to people living here? Could it happen again?

Prepare the questions that you will need to ask when you finally land. Think of all the questions that your readers will want answers to. Questions that begin with *When . . .*, *Where. . .*, *What . . .*, *Who . . .*, *Why . . .* and *How . . .* will provide more information.

You can use all the information in this Building Block to answer your questions about the floods around York. But, be careful! Some of the information may not be relevant to answer your questions (remember that your report is only 500 words). Alternatively, you may not obtain all the information you need, in which case you can try the websites that are mentioned to do further research.

You can use the writing menu on page 104 to help you to write your report.

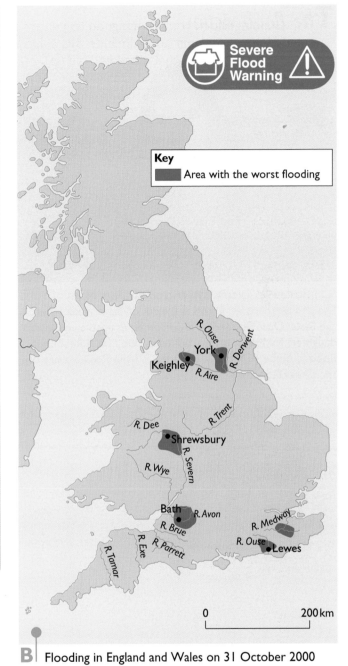

B | Flooding in England and Wales on 31 October 2000

The floods that hit England and Wales in autumn 2000 were the worst for 50 years. From map B you can see that York was not the only area that was flooded. On 30 October a severe storm hit the country, killing six people and causing millions of pounds' worth of damage. Torrential rain and 150 km per hour winds uprooted trees, blocked roads and cut electricity supplies.

The storm left floods in its wake. Many rivers burst their banks and flooded hundreds of homes. The **Environment Agency** – the government body with responsibility for flooding – issued 25 severe flood warnings. This means that there is imminent danger to life and property. The severe weather came after record levels of rainfall in southern England in October and an exceptionally wet September. Throughout the country, the **water table** remained high for the rest of the winter and some land was under water for six months. The twelve months from April 2000 to April 2001 were the wettest on record.

The flood victim

C Rosie Hick lives in Ryther, a small village south of York. She and her partner, David, were cut off for a week when the River Ouse flooded. A few kilometres upstream David's daughter, Kathryn, was marooned on the first floor of her flooded home in Naburn. Extract E is from Rosie's diary of the Great Flood.

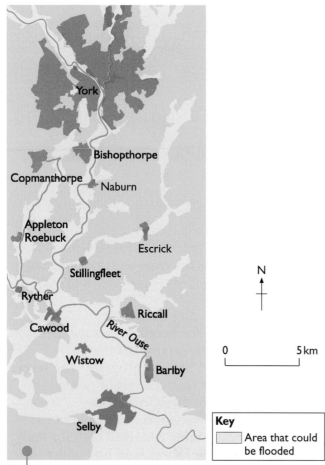

Key
Area that could be flooded

D Flood-risk map of the York area, produced by the Environment Agency. Blue areas are those at most risk from flooding. These areas were flooded in November 2000.

Monday 30 October   Wild day, cold, blowing a gale and snowing sideways. Decorators here to finish glossing bedroom windows, which have to be left open to dry – snow blowing in. Grey, soggy, cold and foreboding.

Tuesday 31 October   Wake to find lanes of water running across fields towards us. River rising alarmingly, barometer dropped to its lowest possible. Kathryn phones: 'Hi Rosie, you and Dad OK? Isn't it great, we're totally cut off!' The joy of youth.

Wednesday 1 November   Decorator finishes and dashes off to escape rising water at front gate. The moggies catch dozens of mice trying to escape from fields.

Thursday 2 November   Begin to feel anxious. Water still rising, five-bar gate is now two-bar gate and we can't get to it. Pumps arrive to clear gardens, garages and drains in village. Helicopters fly overhead on a regular basis. I've a sinking feeling. Keep on Radio York, it's a lifeline – helplines, river reports, sandbags, evacuation. Is it a war zone? Army, police, Environment Agency – it's beginning to really worry me, do they know we are cut off? I feel disorientated; the land's now the sea.

Friday 3 November   Still raining, still rising. Electricity off: light open fires, must keep warm. More floods on TV – Alan Titchmarsh advises not to walk on lawns to clear leaves. Well, Alan, our leaves float!

Saturday 4 November   At last rains stop. My spirits are lifted, then drop when I hear forecast. Feel a bit desperate. We get out the inflatable canoe and go up to village. Back home, septic tank disappeared under water, drains are backing up to house and rising. Can't use toilets, sinks or baths – find portable camping loo in garage, put in bath in case it leaks. Charming. Kathryn phones: 'We got to Tesco by ferry!'

Sunday 5 November   We won't be lighting our bonfire, it's in the new lake. Radio forecasts heavy prolonged rain – no end in sight.

Monday 6 November   Grey, cold and pouring down. Despondent. Garage floods. Five-bar gate gone, totally submerged, must be over four feet [120 cm] deep in inky black, scary stuff.

Tuesday 7 November   Pouring. No refuse collection, burning what we can. Try to walk to Cawood, reach Button Hill – no chance, must be three feet [1 m deep]. Home to see army evacuate elderly neighbour minutes before Button Hill totally impassable. Dark outside, lights blown, quiet except for 24-hour roar of pumps and driving rain. Feel very alone, only the six homes together, totally cut off.

E Rosie's diary of the Great Flood

The weather

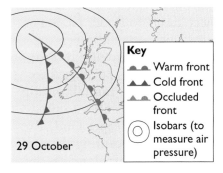

29 October

Key
- 🔵🔵 Warm front
- ▲▲ Cold front
- 🔵▲ Occluded front
- ◎ Isobars (to measure air pressure)

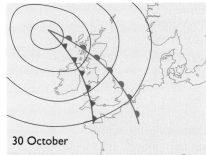

30 October

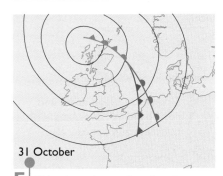

31 October

F Weather charts showing the passage of a **depression** over the UK, 29–31 October 2000

G Satellite photo of a depression over the UK on 30 October 2000

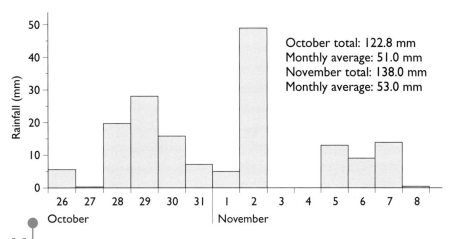

October total: 122.8 mm
Monthly average: 51.0 mm
November total: 138.0 mm
Monthly average: 53.0 mm

H Rainfall at Leeming, in North Yorkshire, October–November 2000. Rainfall on 2 November was almost the same as the total average monthly rainfall for November.
Source: The Met Office. You can obtain more weather data at the Met Office website at www.metoffice.com.

Activities

1 🔲 Read Rosie's diary in E.
  a) First, look at the language she uses. How effective is it? List the words and phrases she uses:
  i) to describe the weather
  ii) to express her feelings.
  You could use some of these words and phrases when you do the assignment at the end of this Building Block.
  b) Now, examine what happened as a result of the floods. List:
  i) the effects that flooding has on people
  ii) the ways that people respond to flooding.

2 Compare Rosie's diary to sources D and F–H.
  a) Locate Ryther on map D. Explain why the village is at risk from flooding.
  b) Look at the passage of the depression in F and G, and the rainfall pattern in H. Explain why the floods happened when they did. (Your teacher may give you more information about the passage of a depression.)

Homework

3 Find out how your local area could be affected by flooding. You could visit the Environment Agency website at www.environment-agency.gov-uk or contact your local council flood advice centre.

The landscape

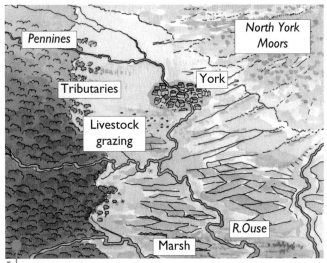

I The natural landscape of the Vale of York, hundreds of years ago

The Vale of York, through which the River Ouse flows, has always been liable to flood. However, in the past the natural landscape did much to reduce the impact of flooding (drawing I). Peat on the moors around the Vale of York used to act as a giant sponge, soaking up rainfall during the winter and slowing down **run-off** into the rivers. Forests, growing on the steeper land, also helped to reduce run-off by intercepting the rain before it reached the ground. Marshland on the **floodplain** surrounding the river acted as a store for the surplus water when the rivers did flood. Drawing J shows how the changes made by people to the landscape have increased both the impact and risk of flooding.

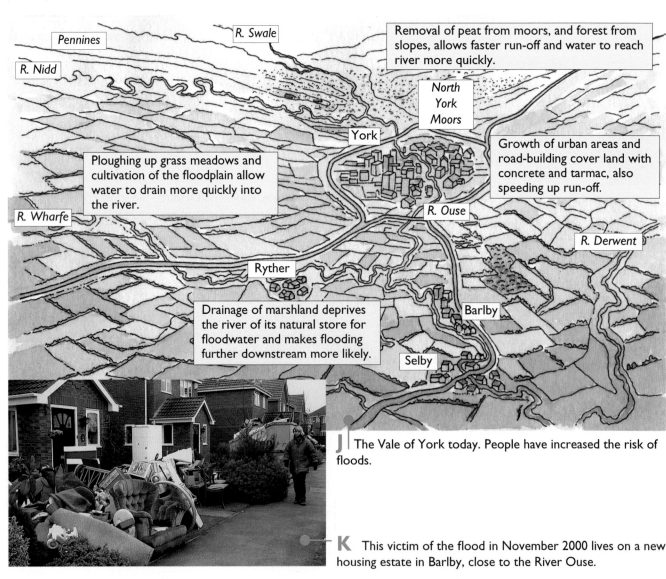

Removal of peat from moors, and forest from slopes, allows faster run-off and water to reach river more quickly.

Growth of urban areas and road-building cover land with concrete and tarmac, also speeding up run-off.

Ploughing up grass meadows and cultivation of the floodplain allow water to drain more quickly into the river.

Drainage of marshland deprives the river of its natural store for floodwater and makes flooding further downstream more likely.

J The Vale of York today. People have increased the risk of floods.

K This victim of the flood in November 2000 lives on a new housing estate in Barlby, close to the River Ouse.

The solutions

For years we have been trying to control rivers to prevent flooding. But the problem still seems to be growing. Some flood prevention schemes actually increase the risk of flooding further downstream. Traditional methods of preventing floods include:

- straightening rivers so that water is able to flow more quickly
- dredging rivers to make them deeper and wider
- building embankments to stop rivers from overflowing
- building dams to control the flow of water in rivers
- and, if all else fails, providing people with sandbags to keep the floodwater out of their homes!

New natural methods of flood control try to reduce the impact of flooding on people. These ideas include:

- allowing rivers to flow freely and to flood in places where they will do least damage
- replacing cultivated land with marsh and meadows that will act as a natural store for floodwater
- planting more trees to reduce run-off
- refusing permission for any more building on floodplains to reduce the danger from flooding.

Continue your investigation about disasters on these two pages.

L The River Ouse flowing near York. How is flooding being prevented?

Activities

1 You are going to draw a large table to show how landscape changes in the Vale of York have increased flooding problems.
   a) Compare drawings I and J. In the first column, list the landscape changes. For example, *Peat is removed from moors.*
   b) In the second column, explain how each change could lead to an increased risk of flooding.
   c) In the third column, explain how easy, or difficult, the changes would be to reverse.

Landscape change	How does it increase flooding?	How easy is it to reverse?
Peat is removed from moors		

2 Look at photo L.
   a) Describe how people have tried to prevent flooding.
   b) Explain why, despite flood prevention measures, this area was flooded again in 2000.
   c) Suggest alternative ways to reduce flooding here.

Assignment

Now that you have studied all the information in this Building Block you can write your newspaper report about flooding around York. Check the instructions for the assignment on pages 98-9. Use the writing menu on page 104 to help you.

When you have written your report, work with a partner. You now have to edit each other's reports. This is what happens in real life when a journalist files a report to their newspaper.

To edit your partner's report:

- cut out words and sentences that are not really needed
- change the order of words and sentences to make it easier to read
- suggest things that could be added to make it more interesting, or easier to understand.

Don't make any more changes than you think are really necessary!

Discuss the changes that you made with your partner and explain why you made them.

BUILDING BLOCKS

How to write a news story

Text type	Tense	Starters	Links	Conclusions	Vocabulary
Recount	*Past*	First, …	… and …	In conclusion, …	hazard
Description	Present	Second, …	… also …	In summary, …	disaster
Method	Future	Next …	… as well as …	Overall, …	flood
Explanation		In the first place, …	Another …	On the whole, …	flood warning
Persuasion		To begin with …	… so …	Today …	Environment Agency
Discussion		Before …	… as a result of …	In future …	depression
		Previously …	… because …	Finally, …	run off
		Later …	… due to …		drainage
		Following …	… however …		flood plain
			… on the other hand …		embankment
			… despite this …		drought
			… mainly …		famine
			… mostly …		aid
			… usually …		desertification
			Unfortunately …		pastoralist
					overgrazing
					debt
					emergency relief
					long-term development

STRUCTURE FOR WRITING A NEWS STORY

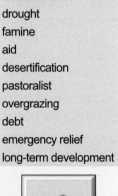

1 Give your story a title. Make it attention grabbing so that people will want to read on (newspaper report), or not switch off (TV report). For example, *Help – we're drowning!* or *Don't give up on Africa*.

2 It is important to have a good starting point for your story to draw the reader/viewer in. This is most likely to be the human angle on the story. If you are writing about floods or droughts you could start with the victim. For example, *On Friday morning Rosie Hick opened the curtains to discover that her garden had become a lake.*

3 Organise the main part of your story chronologically (in the order in which things happened). Broadly, it could fall into three main sections – before, during and after the central event. The first section will deal with the causes of the event. Section two will deal with the event and the immediate impact on people's lives. The third section will deal with people's response and what might happen in future.

   If you are preparing a TV report, use storyboards to sequence the story. Each one will be a single image with supporting script. The script itself is likely to be shorter since some of the information will be in the visual image.

4 You can conclude your story by bringing it back to the starting point. If you had a human angle you could update the reader about what has happened to the person now. For example, *As she hangs the carpet out to dry, Rosie is not worried about another flood. 'We're moving' she explains. 'I could never live through that again.'*

GROUNDWORK

FRAMEWORK

BUILDING BLOCKS

DIGGING DEEPER

In this Building Block you will prepare a TV report about famine in Africa. Use the evidence to continue your investigation about whether disasters are natural, or if people are to blame.

5.6 Why does Africa need aid?

> Now is not the time to give up on Africa. The continent is at an important turning point in its history with some brilliant people trying to make things work better. But sometimes a little help can make all the difference.

A Lenny Henry, one of the celebrities who supports Red Nose Day in aid of Comic Relief.

Every two years the charity Comic Relief organises Red Nose Day. Most schools in the country take part. Yours is probably one of them. But, apart from being a lot of fun and a chance not to wear school uniform for a day, what does it achieve?

Red Nose Day 2001 raised £53.7 million. About two-thirds of this money went to Africa. Twenty of the poorest countries in the world are in Africa and more than a billion people there live in desperate poverty. Many rely on international **aid** to provide them with clean drinking water, healthcare and education. The money raised by Comic Relief is a drop in the ocean when compared to the problems faced in these countries. So can these problems ever be overcome? Lenny Henry believes they can (see photo A). You can find out more about the way that Comic Relief helps Africa at its website www.comicrelief.com.

Assignment

You have been asked to produce a three-minute TV report – roughly 250 words – for Red Nose Day about famine in Africa. The aim is to tell people in the UK about the problems and persuade them to give money.

First, think about the times that you have watched Comic Relief on TV. Usually, the reporter is on location in an African country. They describe a problem that the country faces and explain what the causes of the problem are. Often, local people talk about their everyday lives. At the end, the reporter urges us to give money that will help to solve the problem and make a difference to people's lives.

Use the information in this Building Block to prepare your report. It may help to think about the sort of questions that people ask when they watch Comic Relief:

- Why does famine happen in Africa?
- Why can't people do more to help themselves?
- How will money help?
- Where will my money go to?

You will probably have other questions of your own. You should try to answer these questions in your report.

Prepare your report as a sequence of storyboards (visual images that will appear on screen), together with the text that the presenter will read. You can choose photos, maps or diagrams from the book or do your own research on the websites mentioned. You can use two writing menus to help you. The writing menu on page 104 will help you to tell the story. The writing menu on page 19 will help you to persuade your viewers to give money.

What are the causes of famine?

Large areas of Africa are suffering from **desertification**, where once productive land is turning into desert (see map B). In part this has been caused by **drought** – particularly during the 1970s and 1980s – but it is also a result of human pressure on the land from population growth. It is possible that global warming may have played a part in creating drought, as you can see in source D.

The region most at risk from desertification is the Sahel, a belt of land that stretches across Africa just south of the Sahara Desert. Here many people are **pastoralists**, depending on cattle for their livelihood. Overgrazing, as population grows and land becomes scarce, has made desertification worse. It has ruined grazing land, killed cattle and left people at risk from famine.

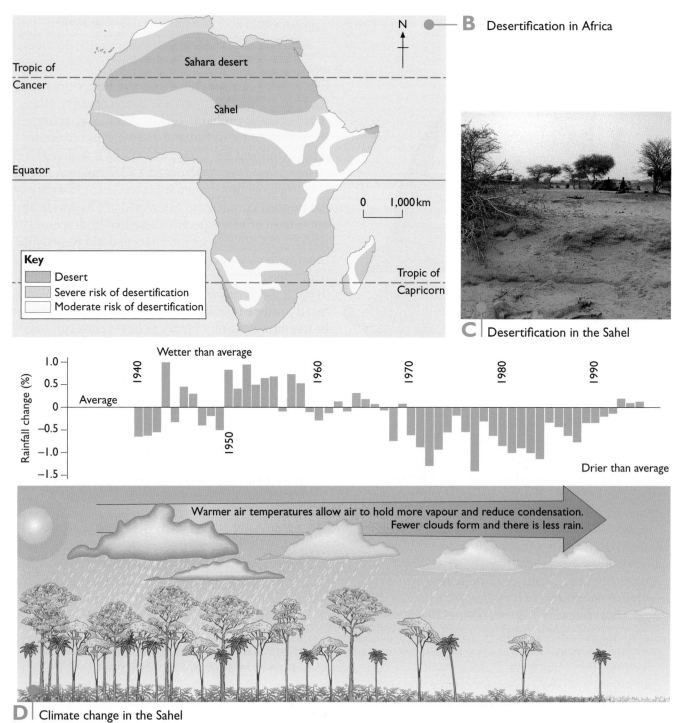

B Desertification in Africa

Tropic of Cancer

Sahara desert

Sahel

Equator

0   1,000 km

Key
Desert
Severe risk of desertification
Moderate risk of desertification

Tropic of Capricorn

C Desertification in the Sahel

Wetter than average

Average

Drier than average

Rainfall change (%)

Warmer air temperatures allow air to hold more vapour and reduce condensation. Fewer clouds form and there is less rain.

D Climate change in the Sahel

Famine also has economic and political causes. Most of the countries in Africa are poor, with low GNP per capita (table E). At the same time, they have high levels of debt as a result of borrowing money in the past. Much of the countries' wealth is used to repay debt. This means people have less money to pay for food and basic services like health and education. In some countries, a large chunk of the money they do spend goes to buy weapons to fight wars. Meanwhile, the best land is used to grow export crops that earn money, not food that people there need to live.

	Ethiopia	UK
Population (millions)	62.4	59.0
Population growth (% per year)	2.7	0.4
Access to safe water (% of population)	27	100
Infant mortality (per 1,000 births)	110	6
GNP per capita ($ per person)	110	20,237
Debt per capita ($ per person)	174	1,400
Debt as % of GNP	158	7
Government spending on health ($ million)	104	80,000

E Ethiopa and the UK compared   Source: Jubilee 2000

Aid from rich countries and international banks is about 10 % of Ethiopia's GNP. Some aid will later have to be repaid.

Military weapons are expensive and use up part of the GNP.

Key
← Income
→ Expenditure

ETHIOPIA

Farming produces almost half of Ethiopia's GNP. Export crops, such as coffee, are the main income earners.

Services, such as health and education, receive very little money because the country is so poor.

Ethiopia's debt is about $10.7 billion – more than its GNP. A large part of the country's income is used to repay debt.

F Income and expenditure in Ethiopia

Activities

1 Study all the information on pages 106–7. Find an example of each of these causes of famine:

> debt   drought   population growth   poverty   export crops   war   desertification   overgrazing   global warming

2 Classify the above factors into human and physical causes. Arrange the words on two sides of a page, human causes on one side, physical causes on the other, with famine in the middle. How many links can you find between them? Include human–human, physical–physical and human–physical links. For example, population growth is one of the reasons for desertification. This is a human–physical link.

Draw arrows to show all the links you find. Label the arrows to explain the links. When you have finished the page could look something like this:

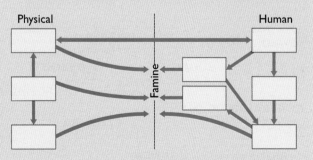

Physical   Human

Famine

Does your diagram help to explain why there is famine in Africa? What do you think is the main cause, or is it impossible to say?

How will money help?

1 April 2000

Famine threat to 12m Africans
Chris McGreal in Johannesburg

A new famine, threatening twelve million lives across seven countries, is looming in the Horn of Africa after long drought and continued conflict, the United Nations warned yesterday.

The spectre is that the scale of the famine that claimed close to one million lives in Ethiopia fifteen years ago, could be repeated. It led to a huge, but belated, international relief operation, led by Bob Geldof's Live Aid concert after the BBC broadcast haunting television pictures of entire villages withering and starving to death.

Once again most of those at risk are in Ethiopia, where aid workers say children are dying by the score from hunger and illnesses related to malnutrition. The Prime Minister, Meles Zenawi, said eight million people faced starvation. In the great drought of 1984–85, almost one million people in Ethiopia were estimated to have died.

The UN's deputy co-ordinator for emergency relief said that unless donors moved quickly to get larger consignments of food on the way, the region could face disaster.

'The rains that have come thus far have been sporadic and uneven, and the prospects for rains in May and June are uncertain,' Carolyn McAskie said. 'We are facing the real prospect in two months from now of another catastrophe which can be averted with the right kind of aid. We all remember the images of suffering of Ethiopians from fifteen years ago and we cannot afford to imagine such a scenario again.'

The UN has appealed for about £126m to feed people across the Horn of Africa. Besides Ethiopia, the other countries at risk are Eritrea, Somalia, Sudan, Kenya, Uganda and Djibouti.

Ethiopia's government blames two years of drought, sporadic heavy rains, frost, black beetle and crop damage by hail for the food crisis. But the war being waged between Ethiopia and Eritrea, the long civil war in Sudan and the clan conflicts in Somalia, have made things worse.

A mother comforts her son at a feeding centre in Ethiopia

G | Adapted from the *Guardian*, 1 April 2000

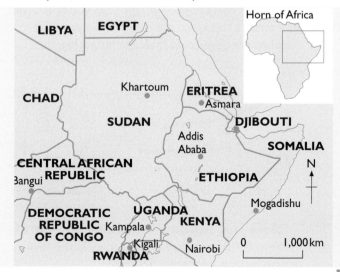

1888–89	Widespread famine
1916–20, 1927–28, 1934–35	Local famines
1969–73	Famine killed 300,000 Ethiopians
1977–78	War and drought in Ethiopia forced one million people to flee into Somalia
1984–85	Famine in Ethiopia killed almost one million people and forced hundreds of thousands to leave their homes
1993	Widespread famine in Somalia caused by civil war
1994	10,000 Ethiopians die in food crisis

H | Famine in the Horn of Africa

Comic Relief raises money to support projects run by **non-governmental organisations** (NGOs) in Africa, as well as here in the UK. Oxfam is a long-established NGO that works in many African countries. It is one of the organisations that Comic Relief supports.

Continue your investigation about disasters on these two pages.

Give a person a fish . . .

Oxfam has been working in Africa for fifty years. It provides **emergency relief** in famines, wars and droughts. In April 2000 it seemed that famine, once again, was about to come to Ethiopia. There had been a three-year drought, crops had failed and livestock were dying from lack of water. Oxfam spent £2 million setting up feeding centres in the town of Gode in eastern Ethiopia. People came to the centres from rural areas throughout the region. The aim was to provide food for people who were malnourished – most likely to be children and the elderly. Medical care was also available and a hygiene programme to reduce the risk of water-borne diseases. Oxfam's efforts reduced the impact of the famine, perhaps saving thousands of lives. Later that month the area had its first rain for three years and the risk of severe famine seemed to have been averted.

. . . and feed them for a day

The grain this woman is carrying will feed a family of six for a month.

A paravet injecting a cow with antibiotics

Teach a person to fish . . .

Oxfam does not just respond to emergencies. When famines, wars and droughts are over, the priority is **long-term development**. The organisation works with pastoralists to help them to adapt their lives to drought in the region. It has become increasingly difficult for people to find water and land for their animals to graze. Oxfam has spent £500,000 on a pastoralist programme in Sudan. The work includes improving water supplies, training paravets to improve animal health and helping local teachers to provide a better education for children. The aim is to enable people to have more control over their own lives.

. . . and feed them for life

Activities

1 What do you think the saying, 'Give a person a fish and feed them for a day, teach a person to fish and feed them for life' means? How does it apply to aid in Africa?

2 Explain why both emergency relief and long-term development are important. You can find out more about Oxfam's work in Africa and elsewhere on their website, www.oxfam.org.uk.

Assignment

Now that you have studied all the information in this Building Block you can prepare your report on drought and famine in Africa for Comic Relief. Check the instructions for the assignment on page 105. Use the writing menus on pages 19 and 104 to help you.

5.7 Global warming – could it be another Titanic?

We started this unit with the story of the *Titanic*. Did you decide that it was a natural disaster, or were people to blame? Floods and droughts are becoming more frequent events in many parts of the world. Does the evidence suggest that these are natural hazards or the result of human activities? In particular, how does global warming contribute to the disasters you have studied?

You can use the story of the *Titanic* as a way to understand what could happen to our planet. Look at source A: are there links between the *Titanic* disaster and global warming?

When the *Titanic* was launched people believed that it was unsinkable. The owners were so confident that they only fitted enough lifeboats for half of the passengers and crew on board.

Icebergs were a frequent hazard on the North Atlantic shipping route. The White Star Line knew this when the *Titanic* set out on its maiden voyage across the Atlantic.

A combination of human errors contributed to the disaster. The sh_ was travelling too fast, the look-ou_ had no binoculars and there were_ enough lifeboats.

Natural hazards, like icebergs, are inevitable but disasters can be avoided. The lessons from the sinking of the *Titanic* were learned and much stricter rules applied to later ships.

The shortage of lifeboats meant that not everybody could be saved. As it turned out, most of the first-class passengers *did* escape while most of the third-class passengers perished.

Large ships take a long time to turn around. Even though the danger was spotted, and a head-on collision averted, the *Titanic* was unable to avoid hitting the iceberg.

A What lessons can we learn from the story of the *Titanic*?

Activities

1 Work with a partner. Look at source A.
  a) Identify the similarities between the sinking of the *Titanic* and global warming. The six labels should give you ideas.
  b) Draw your own labelled diagram of the Earth like the one on the right. Show how global warming could be a disaster in the making.
  c) Are you optimistic or pessimistic that we can avert potential disasters from global warming? Give reasons.

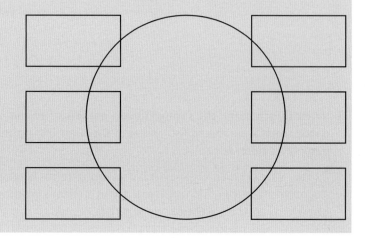

B Two scenes from the film *Titanic*. On the left, steerage passengers remain on the sinking ship. On the right, first-class passengers are lowered in a lifeboat.

	On board *Titanic*	Survivors
Cabin passengers (first- and second-class)	606 (31 children)	381 (31)
Steerage passengers (third-class)	710 (78 children)	270 (26)
Crew	885	52
Total	2,201	703

C Passengers and survivors from the *Titanic*

2 Compare the impact of flooding in the UK and drought in Africa. Look back at the two Building Blocks to do this.
Draw a table to compare:

* the number of deaths
* the area affected
* the duration of the disaster
* the impact on people's lives
* damage to property and the economy
* the response to the disaster.

3 In what ways could the consequences of global warming be more serious in LEDCs than in MEDCs? You can do more research to find out what the impact of global warming could be in different parts of the world. Try these internet websites:
www.epa.gov (US Environmental Protection Agency)
www.climatehotmap.org
http://globalwarming.enviroweb.org

One of the most striking things about the film *Titanic* was the different fates of the rich and poor passengers (photos B). Here the film was based on fact. Data compiled from lists of passengers and survivors shows that it was mainly the rich who survived and the poor who perished (table C). Even though everyone was literally in the same boat, they were unequal in death as they were in life.

The world's population could be compared to the passengers on the *Titanic*. The majority of the world's population (80 per cent) are poor people, living mainly in **less economically developed countries** (LEDCs). A minority of the world's population (20 per cent) are rich people, living mainly in **more economically developed countries** (MEDCs), such as the UK. It is likely that the consequences of global warming will be much more serious for people in LEDCs than for those in MEDCs. Why do you think this is? The two case-studies in this unit – floods in the UK and drought in Africa – should give you some ideas.

5.8
Are these disasters natural?

Assignment

You have been asked by the United Nations to write a report about the increase in the number of 'natural disasters' around the world. They want to know whether or not this is the result of human activities.

To write your report, use the evidence that you have collected throughout the unit for the assignment on page 91. You had to record evidence that disasters are natural or that people are to blame. The images below will remind you of the disasters that you have studied.

Now, pull together all the evidence. You will be expected to write a balanced report that gives evidence both for and against disasters being natural. Look back at page 63 to remind yourself how to write about a controversial issue.

At the end of the report draw your own conclusions based on the evidence, and give your opinion to the United Nations.

Key

| -3 | -2 | -1 | -0.5 | -0.25 | 0 | 0.25 | 0.5 | 1 | 2 | 3 mm/day |

The smooth line on the graph shows the upward trend in the average global temperature

Temperature change (°C)

How much do you know about China? Or should that be how little do you know? China remains a mystery for most of us in the UK. Yet, one-fifth of the world's population lives there. In the twenty-first century it is set to rival the USA as a world superpower. Isn't it time we got to know more about China?

- Do you recognise any of these photos of China? What do you know about them?
- Which comes closest to your image of China?
- Do any of them not fit your image? Why?
- What do the photos tell you about China's past, its people, its power and its potential? (You will come back to this on page 121.)

GROUNDWORK

6.1 Chinese for beginners (and a bit of geography too!)

How many Chinese cities can you name? One, two? If you can manage three you are doing better than average! What about other geographical features – mountains, rivers, deserts? Probably no better.

Compare your knowledge of China with what you know about the USA – a country about the same size as China and with far fewer people. Why do we know so much more about the USA?

One excuse we have for knowing less about China is that the language is so different from ours. Well, you can't use that excuse in this unit! You are about to do a crash course in Chinese, using map A and the dictionary in C.

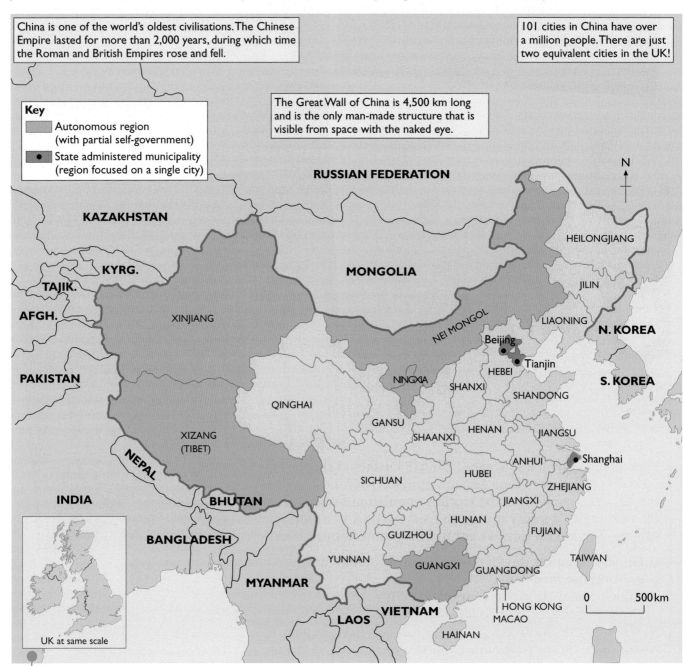

China is one of the world's oldest civilisations. The Chinese Empire lasted for more than 2,000 years, during which time the Roman and British Empires rose and fell.

101 cities in China have over a million people. There are just two equivalent cities in the UK!

The Great Wall of China is 4,500 km long and is the only man-made structure that is visible from space with the naked eye.

Key
- Autonomous region (with partial self-government)
- State administered municipality (region focused on a single city)

RUSSIAN FEDERATION

KAZAKHSTAN
KYRG.
TAJIK.
AFGH.
PAKISTAN
INDIA
NEPAL
BHUTAN
BANGLADESH
MYANMAR
LAOS
VIETNAM

MONGOLIA
HEILONGJIANG
JILIN
LIAONING
N. KOREA
S. KOREA

XINJIANG
NEI MONGOL
NINGXIA
Beijing
Tianjin
HEBEI
SHANXI
SHANDONG
QINGHAI
GANSU
SHAANXI
HENAN
JIANGSU
XIZANG (TIBET)
ANHUI
Shanghai
HUBEI
ZHEJIANG
SICHUAN
JIANGXI
HUNAN
FUJIAN
GUIZHOU
YUNNAN
GUANGXI
GUANGDONG
TAIWAN
HONG KONG
MACAO
HAINAN

N

0   500km

UK at same scale

A China, divided into its provinces

There *is* something Chinese that you almost certainly know about – food! Every town in the UK has a Chinese restaurant and most of us have tasted chow mein, chop suey or spring rolls. But have you ever wondered why there are so many Chinese restaurants in this country? That is a question that you can think about as you study this unit. How many clues can you find?

B | A Chinese restaurant in London

Chinese geographical dictionary

Chinese place names are descriptive. Each syllable has its own meaning. Put the syllables together to find out what the place name means. For example, *shan* is mountain, *dong* is east, so *Shandong* means mountain in the east.

Direction

north	*bei*
south	*nan*
east	*dong*
west	*xi*
back	*hou*
front	*qian*
central	*zhong*
inner	*nei*
outer	*wai*

Features

river	*he, jiang*
ocean, sea	*hai*
gulf, bay	*wan*
lake	*chi, hu*
island	*dao*
mountain	*shan*
valley	*gu*
pass	*guan*
forest	*lin*
cloud	*yu, yun*

Character

peace	*an*
great	*da, tai*
broad	*guang*
long	*chang*
dragon	*long*
city	*shi*
place/region	*zhou*

Colour

black	*hei*
green	*lu, qing*
red	*hong*
yellow	*huang*

Pronunciation

q – ch (as in chin)   x – sh (as in she)   zh – j (as in joke)

C Geography in Chinese!

Activities

1 Test your knowledge of China and the USA. How many of these can you name in each country:
- cities
- rivers
- mountain ranges and deserts
- famous people?

Award yourself one point for each correct answer. How many points do you score for each country? Test yourself again at the end of this unit.

2 Why do we know more about the USA than we do about China? Suggest at least three reasons.

3 Look at map A. Use the dictionary in C to translate the names of the Chinese provinces into English.
  a) On a copy of the map write the English translation for as many provinces as you can. How could this help you to learn about the geography of China?
  b) Make up Chinese names for places in the UK. Draw a local map or a UK map using Chinese names.

4 Why do you think that there are so many Chinese restaurants in the UK? Write a hypothesis to explain this. As you study the rest of this unit look out for clues that you can use to test your hypothesis. These are highlighted as *blue* text. The first one is obvious; *Most people in the UK like Chinese food!*

   Put all the clues together to explain why there are so many Chinese restaurants in the UK.

6.2 Extreme China

China is a country of environmental extremes. To the west is the rugged Tibetan Plateau rising up to the Himalayan peaks at over 8,000 metres. To the east are low-lying, fertile river valleys, plains and deltas. In the north, low rainfall and harsh winters have created an inhospitable desert where life struggles to survive (photo A). In the south, warm temperatures and heavy rainfall enable lush tropical vegetation to thrive (photo B).

Throughout China, people have learnt to live with these physical extremes. From the nomadic herders of the north-west to the rice farmers of the south-east, people have adapted to their environment. It is the physical environment that provides the clue to China's uneven **population distribution**.

A Desert in north-west China

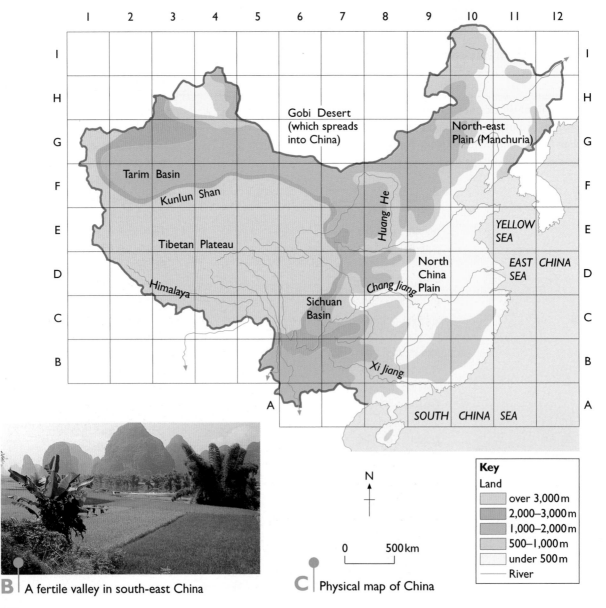

Gobi Desert (which spreads into China)

North-east Plain (Manchuria)

Tarim Basin

Kunlun Shan

Tibetan Plateau

Himalaya

Huang He

YELLOW SEA

North China Plain

Chang Jiang

EAST CHINA SEA

Sichuan Basin

Xi Jiang

SOUTH CHINA SEA

N

0   500km

Key

Land

	over 3,000 m
	2,000–3,000 m
	1,000–2,000 m
	500–1,000 m
	under 500 m
	River

B A fertile valley in south-east China

C Physical map of China

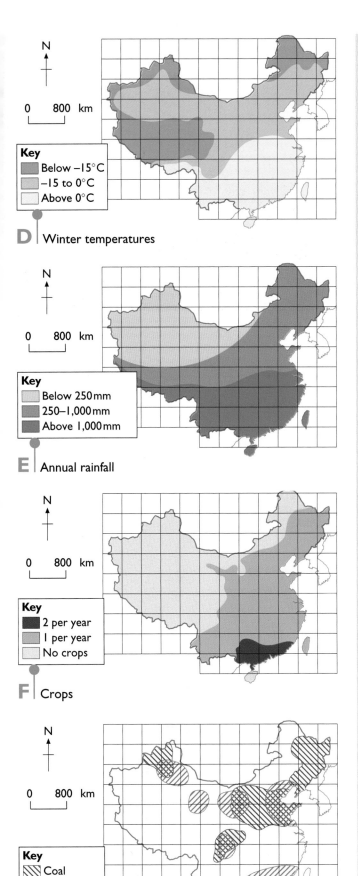

D Winter temperatures

Key
- Below −15°C
- −15 to 0°C
- Above 0°C

E Annual rainfall

Key
- Below 250mm
- 250–1,000mm
- Above 1,000mm

F Crops

Key
- 2 per year
- 1 per year
- No crops

G Resources

Key
- Coal
- Oil

0   800   km

Activity

N You are going to use the maps on pages 116–17 to predict where people live in China, then you will make your own map of population distribution.

a) Look at maps C–G. Between them they tell you a lot about the physical environment of China. Where would people be most likely to live?

b) Complete a copy of this chart giving two points to the areas where people are most likely to live, one point for areas where they are quite likely to live and none for areas where they are least likely to live. Tick one box in each row.

		2	1	0
Map C	Low fertile land			
	Land 500–3,000 m			
	Land over 3,000 m			
Map D	Winter temp below −15°C			
	Winter temp −15–0°C			
	Winter temp above 0°C			
Map E	Annual rainfall below 250 mm			
	Annual rainfall 250–1,000 mm			
	Annual rainfall above 1,000 mm			
Map F	Two crops per year			
	One crop per year			
	No crops			
Map G	No resources			
	Coal or oil			
	Coal and oil			

c) Maps C–G are divided into grid squares. Your teacher will give you a grid map of China with the same grid squares as the maps. Compare the outline map with each of maps C–G in turn. Write the scores that you would give each square in pencil on the map. Each square on your map should have five scores of 0, 1 or 2. Add the scores that you write in each square.

d) Shade your map to show the number of points in each square. The more points the darker the shade you should use. So, for example,

    0–1 white   2–3 pink   4–5 light red
    6–7 dark red   8+ purple

The shade you use shows the likely population density in each square. The darker the shade the higher the density. The map is a prediction of the population distribution in China.

FRAMEWORK

1.3 *billion people . . .*

China has a population of 1,300,000,000 – more people than any other country. Yet, with 21 per cent of the world's population, China only occupies 7 per cent of the Earth's land surface. Ninety-five per cent of that population lives on less than half of the land so you can see that some parts of China must have a bit of a space problem! How does the population distribution in map B compare with the map that you drew for the activity on page 117?

A Crowds in Beijing

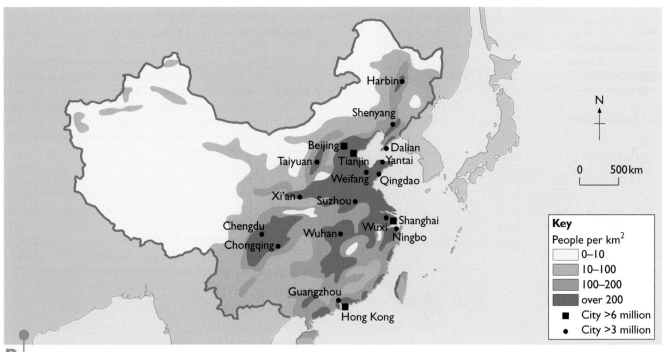

B Population distribution in China

Key
People per km²
- [ ] 0–10
- 10–100
- 100–200
- over 200
- ■ City >6 million
- ● City >3 million

Activities

1 Compare the predicted population distribution map you drew in the activity on page 117 with map B. How accurate was your map?

   Carefully describe the population distribution in China using map B. In your description mention each of these areas: the coastal strip, the North China Plain, inland river valleys, the Tibetan Plateau, the Gobi Desert. You will find these areas on map C on page 116.

2 Explain the population distribution in China. You will need to look back at maps C–G on pages 116–17. Write four short paragraphs to explain the influence on population of:
   i) relief
   ii) climate
   iii) farming
   iv) resources.

... *and counting*

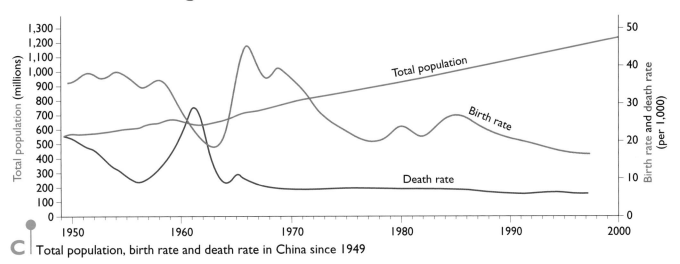

C | Total population, birth rate and death rate in China since 1949

China's population grew throughout the twentieth century. By the 1970s it was clear that the country's **resources** could not continue to support its growing population forever. The government decided to impose a strict **one-child policy** to limit population growth.

The government believed that reducing population growth was more important than the right of individual couples to have children. Today there are 300 million fewer Chinese people than there would have been without the one-child policy, so the government can claim success.

The policy has worked best in urban areas where people have fewer children anyway. It is more difficult to impose in rural areas where 70 per cent of China's population still lives. People here need children to look after them in old age, especially if the first child is a girl. In traditional Chinese communities girls go to live with their husband's family when they get married. If the girl was an only child this would leave her parents with no one to look after them.

China has now relaxed the one-child policy. If two only-children get married they are allowed to have two children of their own. The country's population looks set to peak at 1.6 billion in about 2050 and will then start to decline.

3 N Look at graph C.

a) Describe the changes since 1949 in the total population, the birth rate and the death rate.

b) How do the changes in birth rate and death rate help to explain the change in total population?

c) Draw an extended graph to show what you would expect to happen to the birth rate, death rate and total population by 2050.

4 Should everyone have the right to have as many children as they want, or should the government have the right to stop them?

Work with a partner. One of you put yourself in the position of the couple in photo D. Think of all the reasons for having children.

D A Chinese farming couple

The other person put yourself in the position of the Chinese government. Think of all the reasons for limiting population growth.

Discuss your ideas with your partner. Listen carefully to what they have to say. Can you put any counter arguments that might help them to change their mind? Try to produce a population policy that the government and farmers can agree on.

FRAMEWORK

6.4 Changing China

The Chinese Empire began over 2,000 years ago. At that time Chinese civilisation was far in advance of Europe. People in China lived in cities and used sophisticated farming and building techniques. They had their own medical knowledge, engineering skills and writing system. Many ideas that we use today came from China – including print and paper, the magnetic compass, grid mapping and gunpowder.

It was the famous Chinese philosopher, Confucius, around 500BC whose ideas formed the basis for Chinese civilisation. He believed that every person, from peasant to emperor, should know their role in society and respect their superiors. This was an important foundation of Chinese society for 2,000 years. Only in the past generation have these basic ideas been questioned and begun to change.

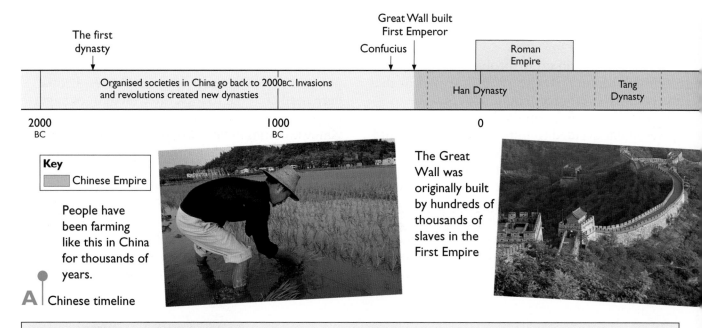

Great Wall built
First Emperor

Confucius

Roman Empire

The first dynasty

Organised societies in China go back to 2000BC. Invasions and revolutions created new dynasties

Han Dynasty

Tang Dynasty

2000 BC   1000 BC   0

Key

Chinese Empire

People have been farming like this in China for thousands of years.

The Great Wall was originally built by hundreds of thousands of slaves in the First Empire

A Chinese timeline

October 1949
Mao Zedong became the first leader of the People's Republic of China. Support for the new Communist Government came from the poorest farmers in the countryside. One of the government's first acts was to take land from rich landlords to give to farmers.

The Great Leap Forward, 1958
The countryside was reorganised to increase production. People's communes were set up in villages to manage farming and industry. They initiated large projects like irrigation and steel-making. There were some successes, but also disasters.

Famine in the early 1960s
Famine was the greatest disaster after the Great Leap Forward. It followed a long drought, but many blamed the famine on the way villages had been reorganised. More

than 40 million people died. It was also a sign that China's population was outgrowing its resources.

The Cultural Revolution, 1966–76
There were disagreements within the Communist Party about its principles and Mao Zedong took more control. Thousands of people whose ideas were not in line with the Party's were sent to prison. Millions of young people were enlisted as 'Red Guards' to spy within their families and communities.

The Open-Door Policy, 1976 onwards
Mao died and the Communist Party was taken over by modernisers led by Deng Xiaoping. They re-established trade links between China and the rest of the world and encouraged foreign companies to come in. This has led to the fastest economic growth in China's history.

B China since 1949

The Empire declined through the nineteenth century at the same time as European involvement in China grew. Britain took Hong Kong on a 99-year lease and finally handed it back in 1997. China was split by internal divisions and the Empire ended in 1912. By this time China had become one of the poorest countries in the world. Later, in 1949, the Communist Party came to power to create the People's Republic of China. It was the beginning of 50 years of upheaval and change in China (see box B).

D The new skyline in Shanghai is an obvious sign of change in China

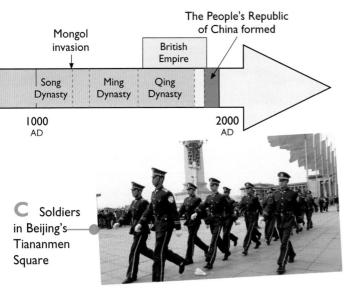

C Soldiers in Beijing's Tiananmen Square

Activities

1 Look at the photos on this page. They are the same photos that you saw on page 113. The information on pages 120–1 should give you more ideas about the photos. You are going to think again about the questions on page 113. What do you now know about these aspects of China: its past, its people, its power and its potential?

Work in a group of four. Each person should choose one of these four aspects to write about. Analyse the photos to describe what they tell you about it. Share your ideas with your group. Is it possible to see the same photos in different ways?

2 Read the information in box B. Compare this information with graph C on page 119. They both cover the period since 1949.

You are going to make a living population graph to show the connection between population change in China and events in the lives of ordinary people. Read the labels below and write them at an appropriate point on a copy of the population graph.

- Zhang Li Ping is forced to have an abortion when she is expecting a second child.
- The population of Jinghong village is halved in one year as people die from hunger.
- Chen Ziming dies, aged 75. At 15 he fought with Mao Zedong when the Communists came to power.
- Lu Ping moves to Guangzhou to work in one of the new factories that has recently opened.
- Wan Enmao has her fourth child after her two youngest children die.
- Yao Shunlan decides to pursue her career as a software consultant rather than have any children.

BUILDING BLOCKS

In this Building Block you will compare the freedom that people have in China with the freedom that we have in the UK. You could do part of your investigation at home!

6.5

How much freedom do Chinese people have?

China has never been a **democracy**. Most Chinese people have spent their whole life under Communist rule and are not allowed to vote. For most of the previous 2,000 years the country was ruled by an all-powerful emperor.

Here, in the UK, we live in one of the world's oldest democracies. We pride ourselves on the freedom that we have and the choices this allows us to make. But what difference does democracy make in people's everyday lives? Do we really have more freedom than people in China? You are going to investigate this question by comparing the lives of an ordinary Chinese family with your own family.

You will meet the Yué family who live in Chengdu, capital of Sichuan, one of China's largest provinces. About 100 million people live in Sichuan. It is roughly the same size as the UK with nearly twice as many people.

Chengdu itself is home to four million people. Its population is rising as people migrate from the countryside to work in new industries in the city. Chengdu is at the centre of the region's oil and gas industry and also has a growing computer industry.

A Yué Wei Sheng (left), Li Xi Feng (right) and their daughter, Yué Ting

Chinese names can seem strange until you get used to them. The family name – equivalent to your surname – always comes first. Yué Wei Sheng's family name is Yué. Children take their family name from their father, like Yué Ting. She is called Ting by her family and friends. Women keep their own family name when they get married. Xi Feng kept her family name – Li – when she married Wei Sheng.

SICHUAN

Chengdu•

B Chengdu city centre

Where can we live?

> Originally I come from Shaanxi, the province north of Sichuan. It used to be difficult to move from one part of China to another. Now companies need qualified workers, so if you have a good education, it is easy to move. Everybody still needs a residence permit to move and to get one you must have a job to go to. We live in an apartment with two bedrooms. This is a typical home in Chengdu. Many apartment blocks have been built in the past twenty years to house all the new workers. We bought our home from a real estate company although it used to be owned by the government. It is easier to move if you own your own home.

Wei Sheng

D Apartment block in Chengdu

What job can I do?

> Both of us work for the Sichuan Petroleum Administration, the government-owned company that drills oil and gas in Sichuan. I work as an English translator and my husband is an accountant. We earn about Y1,500 (£125) each a month. This is an average salary in Chengdu.
> We have both worked for the company for 27 years. There are not many opportunities for promotion and it would be difficult to move to another job. But things are changing. Our daughter, Ting, should have more opportunities than we did. There is more competition for jobs these days. The rewards for hard work are greater, but there is also more unemployment.

Xi Feng

C

Activities

1 Divide a page in your workbook into two halves for China and the UK. You are going to make notes about the freedom that people have in both countries.

2 ⬛ Read the information on this page. Make notes about freedom in China.
   a) Note the freedom that people have about where to live. What limits their freedom?
   b) Note the freedom that people have about what job to do. What limits their freedom?

Homework

3 Interview your own family, or another family you know. Alternatively, you could interview one of your teachers. Ask them about their freedom to choose:
   a) where to live
   b) what job to do.
   What limits their freedom in each case?
   Think about what questions you need to ask. If you read what the Yué family say in source C it should give you ideas.
   Make notes from the interview that you can compare with your notes about the Yué family later, when you do the assignment.

BUILDING BLOCKS

How many children can we have?

The government only allowed us to have one child. If we'd had any more we would have been fined and could have lost our right to a house or to free education.
Most people understand that the policy is for the good of China. It means that we do not have the economic burden of such a large population. It also means that families can afford to give their child a better upbringing and education.

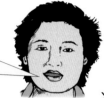
Xi Feng

The problem with an only child is that it can become spoiled and self-centred. We call them little emperors or little empresses. They can grow up without a sense of social responsibility that comes from being part of a family. In any case, the policy only works in cities. People in the countryside don't have any insurance or pensions, so they need children to look after them. The government has begun to relax this policy now.

Wei Sheng

What education will I get?

I am eighteen and next year I hope to go to university to study economics. My parents have paid towards my education since I was six. The more that parents are prepared to pay, the better the school that the child can go to. It gets more expensive when I go to university. They will have to pay Y10,000 a year for my accommodation and Y5,000 for tuition. You can see why it was sensible to have only one child!
When I have got my degree I would like to be able to travel to the USA and get a well-paid job. Many Chinese people want to live and work abroad because the quality of life and the environment are better than in China. This is possible only if you have a good education.

E Ting

Activities

1 ⎡L⎤ Continue to make notes about freedom in China.

a) Note the freedom that people have about a family. What limits their freedom?

b) Note the freedom that people have about education. What limits their freedom?

Homework

2 Continue the interview with your family, or the person that you interviewed on page 123. Ask them about their freedom to choose:

a) whether to have a family

b) education for their children.

What limits their freedom?

Think about the questions that you need to ask.

Makes notes based on the interview that you do.

3 ⎡N⎤ Look at table F.

a) Find out how much an ordinary family in the UK spends each month on these items.

b) Draw a table to compare monthly expenditure for a family in China and the UK.

4 Look at sources H and I.

a) Describe what the graphs in I tell you about:
  i) China's share of global cigarette consumption
  ii) changes in global cigarette consumption.

b) How could advertising like that in photo H be related to what you see in the graphs?

c) Does advertising increase, or reduce, people's freedom? Give reasons for your answer.

How do we spend our money?

Living standards have improved a lot in China during my life. Most people in Chengdu have modern appliances in their home like colour TV, a music system and a fridge. We would like to afford a car but at the moment we just have bicycles.

One of my favourite pastimes is going shopping. In Chengdu we now have modern shopping malls with a wide range of imported goods, as well as Chinese goods. My parents look back to when they were my age, when none of this choice was available.

Ting

	Monthly expenditure (yuan)
House rental (for families who don't own their home)	Y40
Domestic services (electricity, gas, water)	Y150
Food	Y550
Telephone/Internet	Y150
School fees	Y250
Car (running costs)	Y150

F Monthly expenditure for a typical Chinese family living in a city (2002: £1 = 12 yuan). How does your family's compare?

G Shopping in a mall in China

H Cigarette advertising in China. Like in the UK, advertising is now everywhere.

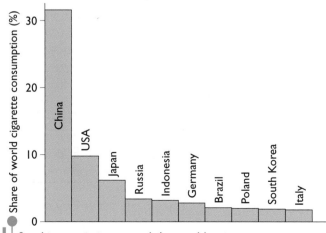

I Smoking statistics around the world.   Source: World Health Organisation, 1996

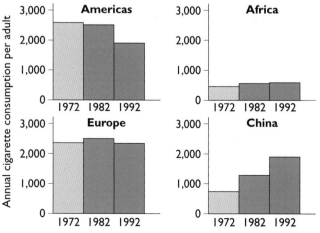

BUILDING BLOCKS

What are we allowed to think and say?

There are two things that the Yué family did not talk about: politics and religion. Chinese people are not really free to talk about what they believe, particularly if they criticise the government. People who protest against the government have been put in jail, or even killed. This happened to protestors in Tiananmen Square in 1989 (look back to page 10 in unit 1).

China is a **one-party state**. This means that there are no political parties, other than the ruling Communist Party. It is illegal to try to set up another party. The Communist Party has atheist principles and does not recognise the existence of God. Although it tolerates religions, it will crack down on them if they criticise the government or try to influence people.

But how important are political and religious freedom? Is the freedom to think or say what you want as important as our other basic freedoms, such as where to live or what job to do? In the UK General Election in 2001, 43 per cent of adults did not vote for any political party. They did not use their political freedom at all. Less than ten per cent of the UK population are active members of any religious organisation. If people in the UK – who do have political and religious freedom – don't use it, should the Chinese government have anything to worry about?

People reading newspapers in Chengdu. Newspapers print only the information that the government wants people to read, and are not allowed to be critical of the government. This is called **censorship**.

Activity

You are going to produce a Human Rights Charter. You can then use it to compare China and the UK in your assignment.

Read the list below of freedoms (or rights) that people could have. Put them into your own order of priority. Are there any others that you want to include? Do you think that any of them should not be included?

People should have the right to:

- choose where to live
- live in a decent house
- be able to work
- choose the job they do
- have children
- get free education for their children
- choose the school their child goes to
- enjoy a decent standard of living
- be able to spend their money as they want
- read any information they want
- vote for the government they want
- follow their own religion
- say what they want.

Use your Human Rights Charter as a checklist to help you to compare China and the UK.

Assignment

You are going to write an account to compare levels of freedom that people have in China and the UK.

Divide your comparison into sections based on the rights in your Human Rights Charter. Read the notes that you have made about the amount of freedom of the Yué family in China and of the family or person you interviewed in the UK. Use your Human Rights Charter as a checklist when you write your comparison. Remember that not all families may have the same amount of freedom within each country. Think how freedom may vary for people who are rich or poor, living in urban or rural areas, male or female.

You can use the writing menu on page 127 to help you.

BUILDING BLOCKS

How to write a comparative account

Text type	Tense	Starters	Links	Conclusions	Vocabulary
Recount	Past	First, …	… compared with …	In conclusion, …	freedom
Description	*Present*	Second, …	In comparison with …	In summary, …	democracy
Method	Future	Next …	Similarly …	To sum up, …	housing
Explanation		In the first place, …	Likewise …	Overall, …	employment
Persuasion		To begin with, …	Equally …	On the whole, …	income
Discussion		Before …	As with …	Finally, …	one-child policy
			… are similar to …		population
			… but …		education
			… yet …		living standards
			… whereas …		expenditure
			… while …		politics
			However, …		religion
			By contrast, …		censorship
			… instead …		

STRUCTURE FOR WRITING A COMPARISON

1 The title should say what you are going to compare in your account. For example, *Comparing levels of freedom in China and the UK.*

2 Write an introduction to the places or features that you are going to compare. If you are comparing two countries, you could mention anything that might affect the comparison that you are making. For example in comparing China and the UK you could mention the size, population and a brief history of each country.

3 Divide your account into sections. Each section should deal with one aspect of the comparison. This is easier than writing everything about one of the places or features and then writing about the other one. For example, *Firstly, let's consider the freedom to choose where to live. In China . . .*

Don't forget that when you compare it is important to describe the similarities as well as the differences. For example, *Likewise, in the UK, people are not always free to choose where they live . . .*

4 Write a conclusion to evaluate the extent of the differences between the places or features that you have compared. Do they differ totally, considerably, slightly, or are they exactly the same? For example, *Overall, the amount of freedom that people in China and the UK have . . .*

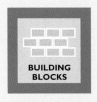

In this Building Block you will investigate energy consumption and production in China. You will write a report about the best strategy for a sustainable future.

6.6

Is China's economic growth sustainable?

China has one of the world's fastest-growing economies. Since 1978 its **gross domestic product** (GDP) has grown by an average of nine per cent per year. It is about four times higher now than it was twenty years ago. However, China still has a long way to go before it is as wealthy as countries in Europe or North America. China's **economic growth** also hides growing inequality within the country. While city dwellers have seen the benefits, living standards for people in the countryside have been much slower to improve.

So, can China continue to raise living standards for all its people? What will be the impact on the environment in China, and on the rest of the world? In other words – is China's economic growth **sustainable**?

A   Zhang Jian Hua lives in Beijing. He and his family live in a villa with all mod cons, including air conditioning for the hot summers and central heating for the cold winters. He also runs a BMW. His job sometimes requires him to fly to other parts of the country rather than take the longer journey by train. He has a high-energy lifestyle.

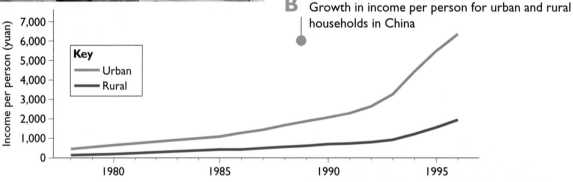

B   Growth in income per person for urban and rural households in China

C   Guo Li Ping owns a small farm. She and her family do all the work by hand as they cannot afford machinery. They use bicycles rather than a car for their journeys. The village where they live has electricity, and homes are equipped with modern appliances. However, they still burn wood for their cooking and heating. Seventy per cent of Chinese people still live in rural areas and most of them have a low-energy lifestyle.

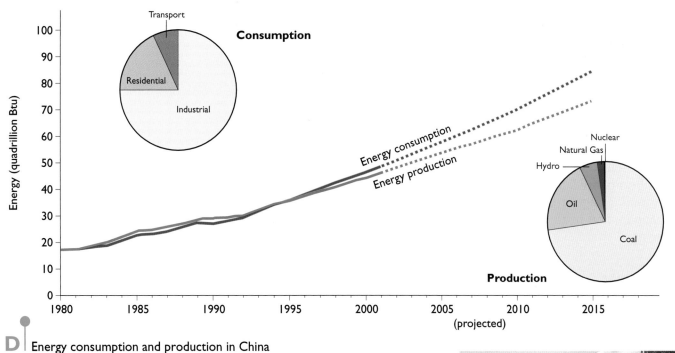

D Energy consumption and production in China

China is the second largest energy consumer in the world, after the USA. It consumes about ten per cent of global energy, but this is rising as the country's economy grows. Already consumption has overtaken production, so that China now has to import oil (graph D). The environmental impact of using energy worries people inside and outside China. Eight of the ten most polluted cities in the world are in China, including Beijing (photo E). And China's emissions of greenhouse gases (see page 96 in unit 5) are a threat to the world's efforts to curb global warming. Currently, the USA is the world's largest producer of greenhouse gases, but China is likely to overtake it by 2015.

E Air pollution in Beijing

Activities

1 You work for the energy department of the Chinese government. You have been asked to predict how energy consumption is likely to grow in China up until 2050. These are the factors that you must take into account.
   - Population growth (see graph C on page 119): population is set to peak in 2050 at 1.6 billion.
   - Economic growth: the government wants to maintain rapid economic growth at or around nine per cent per year (see graph B opposite).
   - Inequality: people in the countryside want to see their living standards improve faster to match those in urban areas (see graph B opposite).
   a) Consider how each of the above factors will affect energy consumption. On a copy of graph D, extend the energy consumption line to the year 2050. What might the level of energy consumption be by then?
   b) Write a paragraph to explain the line that you have drawn.
   c) What other factors might you need to know to make an accurate prediction?

2 The average person in the USA consumes ten times the energy of an average Chinese person. US President George W. Bush said that the USA would reduce its emissions of greenhouse gases only if China does the same.
   a) Do you think that he has a right to demand this? Give reasons.
   b) What response do you think that the Chinese President might give to the US President?

What are China's energy options?

China has had to look at other sources of energy to meet the growing demand. It also wants to reduce the country's dependence on coal. The most ambitious scheme of all is the Three Gorges Dam Project on the Chang Jiang (Yangtze River). When it is complete in 2010 it will be the largest dam in the world, generating 18 million kW of **hydro-electricity**, enough to supply a city of ten million people. The vast scale of the project has caused a huge controversy. Source G shows the benefits and problems that the dam is likely to bring.

F One of the gorges on the Chang Jiang

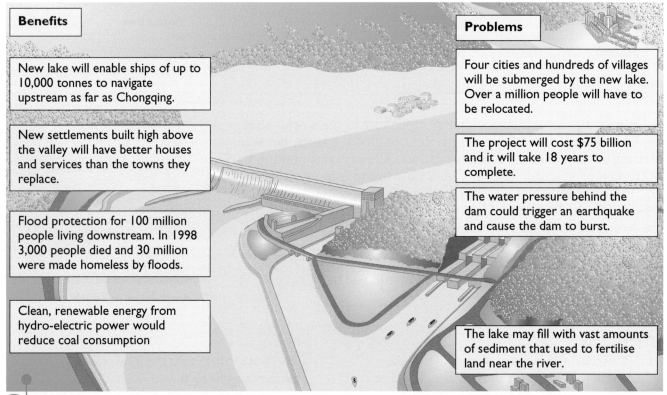

Benefits

New lake will enable ships of up to 10,000 tonnes to navigate upstream as far as Chongqing.

New settlements built high above the valley will have better houses and services than the towns they replace.

Flood protection for 100 million people living downstream. In 1998 3,000 people died and 30 million were made homeless by floods.

Clean, renewable energy from hydro-electric power would reduce coal consumption

Problems

Four cities and hundreds of villages will be submerged by the new lake. Over a million people will have to be relocated.

The project will cost $75 billion and it will take 18 years to complete.

The water pressure behind the dam could trigger an earthquake and cause the dam to burst.

The lake may fill with vast amounts of sediment that used to fertilise land near the river.

G The Three Gorges Dam

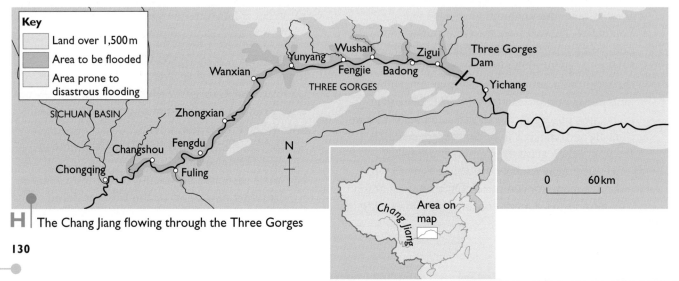

Key

☐ Land over 1,500 m
☐ Area to be flooded
☐ Area prone to disastrous flooding

SICHUAN BASIN

Wanxian

Yunyang Wushan Zigui Three Gorges Dam
Fengjie Badong Yichang
THREE GORGES

Zhongxian

Fengdu

Changshou

Chongqing Fuling

N

Chang Jiang Area on map

0 60 km

H The Chang Jiang flowing through the Three Gorges

One-fifth of China's population – mainly in remote rural areas – has no electricity supply. The distribution of energy across such a vast, mountainous country is a real problem. An alternative to large scale energy projects like the Three Gorges Dam is many small projects to meet the energy needs of each village.

One method that is already used by millions of rural inhabitants is the **biogas** generator (drawing I). This converts human and animal waste into methane gas that can be used for heating, cooking or as fuel for machinery. Biogas is one of many types of **renewable energy** that could help to meet China's demand for energy. Small-scale hydro-electric schemes, wind power and solar power are all used in China and could become more important.

Another priority is to improve energy efficiency and reduce waste. There are plans to reduce coal production and use more oil and gas. Coal produces more air pollution than other types of fuel. It is also more difficult to transport and this can waste energy even before the coal is burnt. As for transport, China could do worse than look again at the bicycle (photo J). At the moment bicycles still outnumber cars in China, but imagine the consequences if it was the other way round!

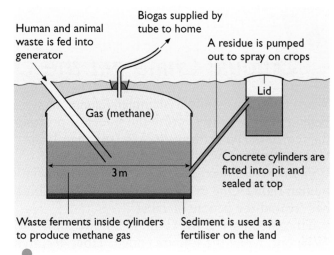

I │ A biogas generator

J │ Bicycles in Chengdu. Most people do shopping by bicycle, but for how much longer?

Assignment

You work for an environmental organisation. You have to produce a report entitled, 'China – towards a sustainable future'. It should suggest how energy use in China could become more sustainable.

Write your report in two sections.

- In the first section, about *energy consumption*, include the graph you drew for activity 1 on page 129. For how long could energy consumption continue to grow like this? How could the growth be slowed down, or even reversed?

- In the second section, about *energy production*, suggest which sources of energy would be most sustainable for China. Think about conventional sources (coal, oil, gas and nuclear power) and alternative sources (wind, solar, wave, geothermal and tidal power). Which sources does China have? Which would have the least harmful impact on the

environment? How long are these sources likely to last?

ICT You can carry out extra research on the environmental impact of different energy sources. Find out about conventional sources of energy and alternative sources of energy. You can do your research on the internet. Some useful websites you could try are:

www.crest.org (Centre for Renewable Energy and Sustainable Technology)
www.nrdc.org (Natural Resources Defence Council)
www.earthdog.com.

Write a conclusion to show how your ideas would be able to balance energy consumption and production, to make energy use in China sustainable.

DIGGING DEEPER

6.1 Should China get the Olympic Games?

When tanks rolled into Tiananmen Square in Beijing in June 1989 to crush the students' protests the rest of the world was shocked (see page 10 in Unit 1). Perhaps, we shouldn't have been. The Chinese government has a reputation for ignoring **human rights**. One example is the way that China has treated the people of Tibet.

Tibet was part of the Chinese Empire until 1912 but, when the Empire ended, Tibet declared independence. However, it was reclaimed by China after 1949. Since then the Chinese government has encouraged millions of Chinese people to move there and they now outnumber Tibetans (see map B).

Tibetans follow the Buddhist faith and have their own culture and language. They do not see themselves as Chinese. During the Cultural Revolution, 5,000 monasteries in Tibet were destroyed and over a million Tibetans died through torture, execution, starvation or in combat against the Chinese. The spiritual leader of the Tibetans – the Dalai Lama – lives in exile and thousands of Tibetans have gone to prison for opposing Chinese rule. But the Chinese government is not willing to give up Tibet. The region has many natural resources, including timber and minerals. China has spent millions of yuan to improve living standards, services and transport in Tibet.

A Tibetan people

B Areas of Tibet settled by Chinese people

Activity

Work with a partner. You are going to play the roles of the Dalai Lama and the Chinese President.

a) Study all the information on this page. If you are being the Dalai Lama, list all the arguments you can find for the independence of Tibet. If you are being the Chinese President, list all the arguments for Tibet to be part of China.

b) Role play a meeting between the two to negotiate the future of Tibet. Discuss all the things that you have written on both your lists.

Are there any areas where you can agree? For example, could China rebuild the monasteries that were destroyed, or could Tibet allow Chinese immigration to continue? Is it possible to find a compromise between independence for Tibet and full integration with China?

c) At the end of the meeting, issue a joint statement to outline the things that you agree and disagree about. Were you able to find a compromise?

The 2008 Olympic Games will be held in Beijing. The Olympics, which is held every four years, is the world's greatest sporting event. It brings together competitors from every nation and is televised worldwide. Hosting the Olympics can bring a country great prestige and huge economic benefits. Some people argue that China should not be allowed to host the Games until it improves its record on human rights. Others think that all the international attention China gets means that it *will have to* improve human rights. Which of them do you think is right? Consider the arguments in table D.

C China's bid for the 2008 Olympics has popular support. Ninety-six per cent of Beijing's population want the Games to be held there.

The carrot approach

Use trade and sporting contacts with the country to influence the government. Try to persuade it to change its policies. The danger is that it will not listen and continue to behave in the same way. There are no international laws that governments have to obey.

The stick approach

Boycott (refuse to do business with) the country. This would isolate it and put pressure on the government to change its policies. However, there would be no opportunity to influence the government. This might also unite the people and government against the rest of the world.

D Two ways to deal with China

Assignment

You are going to take part in a class debate on the motion, 'The Olympic Games should not be held in China unless it respects human rights.'

Decide which side you will take in the debate. Think about all the things that you have learnt about China in this unit. Write a short speech for or against the motion (the writing menu on page 19 will help you). Listen carefully to all the arguments on both sides of the debate. Take a class vote at the end of the debate. If it were up to your class would the Olympic Games be held in China?

After the debate, write a letter to the International Olympic Committee to say whether you agree, or disagree, with the decision that the games will go to China. What advice would you give them about the issue of human rights and how they deal with the Chinese government?

Extra

1 Look back at the first activity that you did in this unit on page 115. You had to test yourself on your knowledge of the USA and China. Test yourself again. How many points do you score for China now?

2 In activity 4 on page 115 you wrote a theory to explain why there are so many Chinese restaurants in the UK. What evidence have you found in this unit? Did it support your theory? Now how would you explain why there are so many Chinese restaurants?

Can we make the world a better place?

Do you ever feel that it's impossible to change the world? After all, what can one person do? I expect the people who built the Great Wall of China felt the same way. But, they didn't know that, one day, people would go into space and that *their* wall would be the only human structure visible on Earth. You *can* change the world!

In this book, I hope you have learnt about geography and citizenship. You must have noticed that the world *is* changing – sometimes for the better, sometimes for the worse. And it is people that make the difference! That is what citizenship is really about. These two pages are to help you to think about what you have learnt and what *you* could do to help make the world a better place.

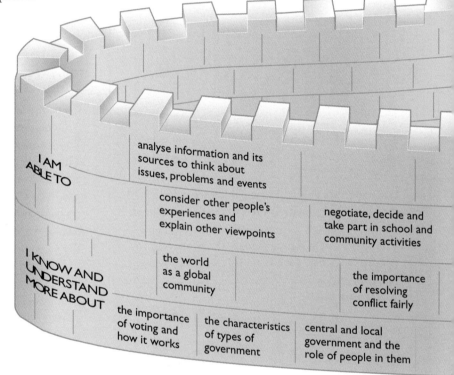

I AM ABLE TO
- analyse information and its sources to think about issues, problems and events
- consider other people's experiences and explain other viewpoints
- negotiate, decide and take part in school and community activities

I KNOW AND UNDERSTAND MORE ABOUT
- the world as a global community
- the importance of resolving conflict fairly
- the importance of voting and how it works
- the characteristics of types of government
- central and local government and the role of people in them

Activities

1 Look at the people on these pages. You have met them before in the units in this book.
  a) Can you remember who they are (names don't matter)?
  b) In which unit did you meet each person?
  c) In what ways do you think each person is helping to change the world (for example, Benjamin Zephaniah is helping people to understand Britain's multicultural society)?
  d) Write one sentence to describe the main thing that you think *you* learnt in each unit.

GREAT
CITIZENSHIP
WALL

reflect on the
process of participating

contribute to
discussions and
take part in debates

justify opinions
orally and in
writing

the work of local,
national and international
voluntary groups

the importance
of the media in
society

legal and
human rights
in society

different groups of people
in the UK, and the need
for mutual respect

2 Look at the Great Citizenship Wall.
Which areas of knowledge and understanding and which skills have
you improved while using this book? For each one, find one piece of
work to demonstrate this.

3 Devise a personal Citizen's Action Plan. Think of the one topic or
issue that you have studied in this book that interests you most.
  a) What more would you like to know? Write a list of questions.
  b) Where could you find out more? Plan some research that you
  could do.
  c) How could you get involved? Find out which organisation you
  could join, or which other people are interested.

Glossary

accessible – easy to get to

advertising – publicity to help a company sell its goods or services

aid – help, in the form of money, food, people or equipment, usually sent by richer countries to poorer ones

asylum – a safe place to live

asylum seeker – a person who wants to be allowed to live in a country where they will be safe

biogas – a renewable source of energy formed from organic waste

border – a line on a map showing the edge of a country

borough – a smaller administrative area within a city

brand – a name, or trademark, for a particular type of goods

by-pass – a road built around one side of a town to reduce the traffic passing through the centre

capital city – the chief city in a country, where the government is

censorship – limitations by government on what people are allowed to read or see in the media

climate – the average weather pattern recorded over many years

commuter – a person who travels to work each day

congestion – too much traffic for it to move freely

conservation – protection of the environment

constituency – an area of the UK represented by its own MP in Parliament

counter-urbanisation – a fall in the proportion of people living in towns and cities

county – an administrative area of a country

deforestation – the destruction of forest

democracy – a system where the government is chosen by the people every few years

deported – forced to leave the country

depression – a moving area of low pressure, which usually brings rain

desertification – the process by which land that was once productive turns to desert

devolution – some powers given by national government to a region, or country, within its borders

dot.com company – a company that sells only through the internet

drought – a long continuous period of dry weather

economic growth – an increase in the amount of goods produced, leading to an increase in wealth

ecosystem – the community of plants and animals in an area and the links between them

emergency relief – aid that is given after a disaster to help people to survive

endangered species – a plant or animal species threatened with extinction

energy – power, in the form of electricity or fossil fuel

Environment Agency – the government organisation that has responsibility for environmental issues, like flooding, in the UK

environmentalist – a person who is concerned about protecting the environment

Environmentally Sensitive Area – an area where the beauty of the landscape depends on maintaining traditional farming methods

euro – the currency of the European Union

European Union – an organisation of European countries which encourages economic links

extinction – the disappearance of a plant or animal species from the planet

factory – a place where goods are manufactured

floodplain – a wide, flat area beside a river, formed by sediment brought downstream when the river is in flood

global economy – the way that goods and services produced in one part of the world are sold and used somewhere else

global warming – the increase in average temperatures that is happening around the world

globalisation – the process by which companies move around the world, helping to create the global economy

GNP per capita – GNP divided by population, as a measure of a country's wealth

government – a group of people who have the power to run a country

green belt – land outside the urban area that is protected from building development

greenhouse effect – the way that gases in the atmosphere trap heat from the Sun

gridlock – the point at which congestion is so bad that traffic comes to a standstill

gross national product (GNP) – the total value of all the goods and services produced by a country

gross-domestic product – the total value of goods and services produce within a country

habitat – the environment in which a plant or animal lives

Heritage Coast – a stretch of beautiful coast that is protected

human rights – basic freedoms that all people should be able to expect, for example not to be persecuted

hydro-electricity – electricity generated by water power

immigrant – a person who has come to live in another country

independence – freedom from government by another country

industry – an economic activity, or type of work

integration – countries becoming closer politically and economically

intensive farming – farming that uses many inputs and changes the environment

internet – global network of computers linked by telephone lines

labour – people who work

less–economically developed countries – poorer countries with a low GNP per capita

local government – a group of people who have the power to run a county or borough

long-term development – improvements in people's quality of life over many years

manufacturing industry – a type of work that makes goods to sell

market – a place where goods are sold or, at a wider scale, the people who buy the goods

mental map – a kind of picture in a person's mind of where places are and what they are like

Meteorological Office – the government organisation that records and predicts weather patterns

metropolitan county – an administrative area based on a major city, or urban area

migrate – move to a new area or country

more-economically developed countries – richer countries with a high GNP per capita

motorway – a major road designed to speed up traffic

MP – a Member of Parliament elected to represent a constituency

national park – a large area of beautiful countryside that is protected by law

nationalism – love of your own country, or the wish for independence

nationality – belonging to a country or from a country

natural hazard – an event that cannot be predicted, creating danger for people

nature reserve – an area where plants and animals need to be protected from human activities

network – a system of connecting routes

newly-industrialising country – a country that is changing from a mainly agricultural economy to a mainly manufacturing economy

non-governmental organisation – a group, independent of any government, that helps to provide aid

one-child policy – the Chinese government policy of limiting the number of children that couples are allowed to one child

one-party state – a country with no democracy where there is only one political party

on-line shopping – buying via the internet

pastoralist – person whose lifestyle depends on keeping animals

pedestrianised area – an area where cars are banned

physical map – a map showing natural features such as rivers and mountain ranges

political map – a map showing artificial areas such as countries and cities that have been created by people

pollution – making the environment unhealthy or unpleasant by adding something

population density – the number of people living in a given area, e.g. 1 km^2

population distribution – the way that people are spread within an area

precipitation – rain, hail and snow

production – making goods from raw materials

public enquiry – a consulting exercise where the government asks concerned groups and individuals for their views about an issue

public transport – buses and trains for people to travel on when they need to

raw material – material (often in its natural state) used to make a product

refugee – a person who has fled their home to live somewhere safer

renewable energy – energy that can be produced from the same source over and over again

resource – something that is useful to people, for example oil, coal, water, food

ring road – a road built all the way around a town to keep traffic out of the centre

run-off – the movement of rainwater over the ground

rural – of or in the countryside

settlement – a place (e.g. town or city) where people live

sustainable – something that can be continued because it is not wasteful

sweatshop – a workshop or factory with poor working conditions

televillage – a village built especially for people who use computers to work at home

teleworker – a person who uses a computer to work at home

theory – an idea that can be tested by using evidence

traffic calming – measures to slow down cars, e.g. with speed bumps and lower speed limits

traffic management – measures to speed up the flow of traffic, e.g. with roundabouts and one-way systems

tranquil areas – rural areas that are some distance from towns, cities, motorways, roads, railways and power stations

trans-national company – a large company that operates in more than one country

urbanisation – an increase in the proportion of people living in towns and cities

water table – the level of water in the ground below which rock is saturated

weather – day-to-day changes in the atmosphere

wilderness – a large natural area untouched by human activities

Index

●

Acknowledgements

The author and publishers are grateful to the following for permission to include material in the text:

p.9 'The British' from *Wicked World*, Benjamin Zephaniah, Puffin Books 2000; p.12 Office for National Statistics © Crown copyright 2000; p.13 Home Office © Crown copyright 2000; Refugee Council Credit to the Nation; p.16 Office for National Statistics © Crown copyright 1997; p.24 Department for Environment, Transport and Regions © Crown copyright 1997 p.27 Redfern, R, *Too many wheels, not enough space*, Wideworld, Vol. 10, Issue 3; p.28 Ben Elton, *Gridlock*, Warner 1992; p.29 Ian Hunt, 'Costing the Air', *Guardian Education*, 25 January 2000; p.31 *Issues in the New Europe*, Graham Drake, Hodder 1994, reproduced by permission of Hodder and Stoughton Educational Limited; p.32 Copyright, Compass Maps Limited; p.38 © Michelin Travel Publications – Extract from GREAT BRITAIN AND IRELAND TOURIST AND MOTORING ATLAS 2002 – Authorisation no. 02020756; p.46 © Fila; © Umbro International Limited; © Reebok; © Le Coq Sportif; p.53 International Labour Office; p.55 The cartographic information which appears in this publication was supplied by © Automobile Association Developments Ltd LIC004/02 [2001] All rights reserved and includes mapping data supplied by Ordnance Survey, Crown Copyright [2001] All rights reserved. Licence 399221; p.56 *Fever Pitch*, Nick Hornby, Penguin 2000; p.60 Maquila Solidarity Network, Canada; p.68 *As I Walked Out One Midsummer Morning*, Laurie Lee, Penguin 1971; p.71 CPRE and the Countryside Commission, October 1995; p.73 John Vincent and Paul Brown, 'Fall in some common bird numbers accelerating', *Guardian*, 6 February 2001; p.74 *Student Atlas*, Collins Longman 1996; p.75 © WWF; p.76 Reproduced from the Ordnance Survey 1:50,000 Landranger mapping with the permission of The Controller of Her Majesty's Stationery Office © Crown copyright. All rights reserved. Licence number 100017789; p.77 Reproduced from the Ordnance Survey 1:50,000 Landranger mapping with the permission of The Controller of Her Majesty's Stationery Office © Crown copyright. All rights reserved. Licence number 100017789; p.79 © Copyright Farming and Countryside Education; p.82 © British Antarctic Survey/ Steve Marshall; p.83 © British Antarctic Survey/ Steve Marshall; p.84 © British Antarctic Survey/ Steve Marshall; p.85 © International Association of Tour Operators; p.88 © British Antarctic Survey/ Steve Marshall; p.92 Information supplied by the Met Office © Crown Copyright; p.94 Information supplied by the Met Office © Crown Copyright; p. 96 Based on information supplied by the Met Office; p.97 Based on information supplied by the Met Office; p.100 Copyright of the Environment Agency; Adaptation of 'Rosie Hicks' Diary', Rosie Hicks, *Guardian*, 11 November 2000; p.101 Based on information supplied by the Met Office; p.108 Adaptation of 'Famine Threat to 12m Africans', Chris McGreal, *Guardian*, 1 April 2000; p.114 Adapted from *China in Change*, David Money, Hodder 1996, Reproduced by permission of Hodder and Stoughton Educational Limited; p.117 Adapted from *China in Change*, David Money, Hodder 1996, Reproduced by permission of Hodder and Stoughton Educational Limited; p.118 Adapted from *China in Change*, David Money, Hodder 1996, Reproduced by permission of Hodder and Stoughton Educational Limited; p.125 Data supplied by the World Health Organization.

Photo acknowledgements

Cover: *tr* Rex Features; *cl* Henryk T Kaiser/Rex Features Ltd, *c* Mark Henley/Impact, *bl* Art Wolfe/Science Photo Library, *bc* NRSC Ltd/ Science Photo Library, *br* Travel Ink, p.1 *t* Martin Bond/ Environmental Images, *b* © Chris Stowers/Panos Pictures; p.iv *t* © Wolfgang Kaehler/Corbis, *c* David Nunuk/Science Photo Library, *b* © Alison Wright/Panos Pictures; p.2 Last Resort; p.3 *tl* Peter Baker/ International Photobank, *tr* Hugh Webster/Stock Scotland, *cl* Collections/Paul Watts, *cr* Sealand Aerial Photography, *bl* National Coal Mining Museum for England, *br* Alton Towers; p.5 Sealand Aerial Photography; p.6 Collections/John D. Beldom; p.7 Peter Baker/International Photobank; p.9 Richard Siker/Rex Features; p.11 Popperfoto/Reuters; p.13 *t* © Crispin Hughes/Photofusion, *b* Exile Images; p.14 *tr* Rex Features, *tl, cr, bl & br* Last Resort; p.15 Topham Picturepoint; p.18 Alex Livesey/Allsport; p.22 *tl* Maximilian Stock Ltd/Science Photo Library, *tr* Collections/Gena Davies, *bl* Michael Crabtree/Topham Picturepoint, *br* The Sun/Rex Features; p.23 Martin Bond/Environmental Images; p.24 Last Resort; p.27 © Trip/C. Kapolka; p.28 Sealand Aerial Photography; p.29 *l* David Townend/Environmental Images, *r* John Widdowson; p.31 *t* Francesco Guidicini/Rex Features, *b* David Townend/Environmental Images; p.32 © Bob Krist/Corbis; p.33 John Widdowson; p.34 *t* John Widdowson, *c* Truus van Gogh/Hollandse Hoogte, *b* Michiel Wijnbergh/Hoolandse Hoogte; p.35 *t* Sealand Aerial Photography, *c & b* Last Resort; p.36 Last Resort; p.39 *t* Last Resort, *b* London Aerial Photo Library; p.40 courtesy Tom Fanning; p.41 courtesy Gerald Kells; p.42 Last Resort; p.43 courtesy Amazon; p.44 © Martin Ellard/Dragon; p.45 © Chris Stowers/Panos Pictures; p.47 John Sibley/Action Images; p.48 *t* Stuart Franklin/Action Images, *b* Stu Forstep/Allsport; p.49 Mike Powell/Allsport; p.52 © Crispin Hughes/ Photofusion; p.54 © getmapping plc 2001; p.56 Offside; p.57 *l* Offside, *r* courtesy Nick Robinson; p.58 courtesy www.arsenal.com; p.59 Last Resort; p.60 © Irene Slegt/Panos; p.61 Juergen Hasenkopf/ Rex; p.64 Andy Lyons/Allsport; p.67 © Wolfgang Kaehler/Corbis; p.68 © The National Gallery, London; p.72 *tl* Warwick Sloss/BBC Wild, *tr* Collections/John & Eliza Forder, *bl & br* John Widdowson; p.73 Dietmar Nill/BBC Wild; p.75 *tl* Anup Shah/BBC Wild, *tr* Lynn M. Stone/BBC Wild, *bl* Richard du Toit/BBC Wild, *br* Douglas Faulkner/Science Photo Library; p.76 *t* courtesy Andy Mason, *b* Sealand Aerial Photography; p.77 *t* courtesy Linda Hall, *b* Collections/Oliver Benn; p.78 *l & r* John Widdowson; p.79 Popperfoto/Reuters; p.80 *l* courtesy Andy Mason, *r* courtesy Linda Hall; p.82 *t* British Antarctic Survey, *b* David Vaughan/Science Photo Library; p.83 British Antarctic Survey; p.84 British Antarctic Survey; p.85 British Antarctic Survey; p.88 Marty Lueders/Environmental Images; p.89 *tl* David Nunuk/Science Photo Library, *tr* SIPA/Rex Features, *c* © Clive Shirley/Panos Pictures, *bl* Sam Ogden/Science Photo Library, *br* Rex Features; p.90 Topham Picturepoint; p.91 Merie W. Wallace/The Kobal Collection; p.92 Rex Features; p.93 Jeremy Hartley/Panos Pictures; p.94 Mary Evans Picture Library; p.95 *t* John Widdowson, *b* Collections/Paul Watts; p.98 London Aerial Photo Library; p.100 © Kippa Matthews; p.101 © University of Dundee; p.102 John Giles/PA Photos; p.103 Collections/Michael George; p.105 Nils Jorgensen/Rex Features; p.106 © Jeremy Hartley/ Panos Pictures; p.107 *l* © Fred Hoogervorst/Panos Pictures, *r* Popperfoto/Reuters; p.108 © Crispin Hughes/Panos Pictures; p.109 *t* © Crispin Hughes, *b* © Toby Adamson; p.111 *l* The Kobal Collection, *r* Merie W. Wallace/The Kobal Collection; p.112 *tl* © Crispin Hughes/Panos Pictures, *tr* London Aerial Photo Library, *bl* © Jeremy Hartley/Panos Pictures, *br* Collections/Paul Watts; p.113 *tl* © Trygve Bølstad/Panos Pictures, *tr* © Alison Wright/Panos Pictures, *bl* © Peter Turnley/Corbis, *br* © Chris Stowers/Panos Pictures; p.115 © Inge Yspeert/Corbis; p.116 *t* © Rob Penn/Axiom, *b* © Gordon D.R. Clements/Axiom; p.118 © Wang Gangfang/Panos Pictures; p.119 © Gordon D.R. Clements/Axiom; p.120 *l* © Trygve Bølstad/ Panos Pictures, *r* © Alison Wright/Panos Pictures; p.121 *l* © Peter Turnley/ Corbis, *r* © Chris Stowers/Panos Pictures; p.122 *both* Alan Widdowson; p.123 © Trygve Bølstad/Panos Pictures; p.124 Alan Widdowson; p.125 *t* Alan Widdowson, *b* © Caroline Penn/Panos Pictures; p.126 Alan Widdowson; p.128 *t* Alan Widdowson, *b* © Gordon D.R. Clements/Axiom; p.129 © D. Tatlow/Panos Pictures; p.130 © R. Jones/Panos Pictures; p.131 Alan Widdowson; p.132 © D. Tatlow/Panos Pictures; p.133 © AFP Photo/Goh Chai Hin; p.134 *t* Richard Siker/Rex Features, *c* © Bob Krist/Corbis, *b* Mike Powell/Allsport; p.135 *tl* Copyright 1995, Worldsat International and J. Knighton/Science Photo Library, *tr* David Vaughan/Science Photo Library, *c* © Toby Adamson, *b* Alan Widdowson.

t = top, b = bottom, l = left, r = right, c = centre